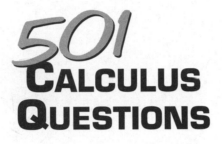

Other Titles of Interest from LearningExpress

501 Algebra Questions
501 Geometry Questions
501 Math Word Problems
1001 Algebra Problems
Algebra Success in 20 Minutes a Day
Calculus Success in 20 Minutes a Day

501 CALCULUS QUESTIONS

Mark A. McKibben, PhD

LEARNINGEXPRESS ®

NEW YORK

ISBN 978-1-57685-765-6

Printed in the United States of America.

9 8 7 6 5 4 3 2 1

For more information or to place an order, contact LearningExpress at:
    2 Rector Street
    26th Floor
    New York, NY 10006

Or visit us at:
    www.learningexpressllc.com

Project Editor: Lindsay Oliver

Contents

About the Author

Dr. Mark McKibben is currently a tenured professor of mathematics and computer science at Goucher College in Baltimore, MD. He earned his PhD in mathematics from Ohio University in 1999, where his area of study was nonlinear analysis and differential equations. His dedication to undergraduate mathematics education has prompted him to write textbooks and more than 20 supplements for courses on algebra, statistics, trigonometry, precalculus, and calculus. He is an active research mathematician who has published more than 25 original research articles, as well as two recent books entitled *Discovering Evolution Equations with Applications, Volume 1: Deterministic Equations* and *Discovering Evaluation Equations with Applications, Volume 2: Stochastic Equations*, published by CRC Press/Chapman-Hall.

Introduction

Why Should I Use This Book?

Simply put, this book will help you achieve success in calculus. When learning calculus, taking time to carefully work through problems is essential. The 501 questions included here, along with the pretest and posttest, will provide plenty of practice opportunities, supplement your study, and help you to master all central single-variable calculus topics.

Many everyday questions involve calculus. For example, if you're trying to calculate the area of an enclosed garden so that you can build a fence, or are trying to optimize the cost of a product or service, or even if you'd like to minimize the amount of material needed to form a box with a certain volume—all of these questions can be answered using calculus. In essence, you may have been *thinking about* and forming *approximate* solutions to calculus problems without ever realizing it. So, calculus is not only an academic topic of study; it also has practical applications to real-life problems.

This book has been designed to provide you with a collection of problems to assist you with single-variable calculus. It has been written with several audiences in mind: anyone who is currently studying calculus and needs additional practice, or someone who has previously taken a calculus course and needs to refresh skills that have become a bit rusty. Instructors teaching a calculus course might also find this collection of problems to be a useful supplement to their own problem sets. Whatever your background or reason for using this book, we hope that you will find it to be a useful resource in your journey through single-variable calculus.

What Is in This Book?

The problems included cover all of the topics typically addressed in courses in single-variable calculus. Multivariable calculus is not included.

Single-variable calculus can be broadly divided into two categories: *differential calculus* and *integral calculus*. Historically, these categories were motivated by the need to address two distinct geometric problems: finding the slope of a tangent line and computing the area of a plane. At the root of both areas is the critical notion of a *limit*. Mastering the rules and techniques outlined in the problem sets in this book will help you to develop an understanding of the central concepts of calculus. In addition, working through these problems will arm you with the tools necessary to attack applied calculus problems accurately and with ease.

How to Use This Book

501 Calculus Questions covers the central calculus topics in 20 chapters. This book also includes a pretest, a posttest, and a handy appendix of calculus formulas and theorems. Before you begin to work the problems in Chapter 1, take the pretest, which will assess your current knowledge of calculus. You'll find the answers in the answer key at the end of the pretest. Taking the pretest will help you determine your strengths and weaknesses in calculus. After taking the pretest, move on to the first chapter.

Chapters 1 through 19 offer questions covering all of single-variable calculus. Be sure to use scratch paper and write down each step as you solve a problem. Remember that most calculus instructors will require you to show your work to receive full credit. Plus, if you skip steps in your answer, this can sometimes cause you to make errors. Use the detailed solutions and answer explanations in the second part of each chapter to check your answers.

Chapter 20 presents common calculus errors. Use this chapter either at the end of your study or as you go along to help you avoid the errors that many people make when studying calculus. When you're ready, take the posttest, which has the same format as the pretest but different questions. Compare your scores to see how much you've improved or to identify areas in which you need additional practice.

Make a Commitment

Success in calculus requires effort. Make a commitment to improving your calculus skills. If you truly want to be successful, make time each day to do problems. When you achieve success in calculus, it will have laid a solid foundation for future academic and real-world successes. So, sharpen your pencil and get ready for the pretest!

Pretest

Before you start Chapter 1, you may want to find out how much you already know about calculus and how much you need to learn. If that's the case, take the pretest, which includes 20 multiple-choice, true/false, and computational questions. While 20 questions can't cover every skill in single-variable calculus, your performance on the pretest will give you a good indication of your strengths and weaknesses.

Take as much time as you need to complete the pretest. When you are finished, check your answers with the answer key at the end of this section.

Questions

1. Which of the following linear equations has a negative slope?
   a. $y = 4x - 1$
   b. $-3 = y - 2x$
   c. $-5x + y = 1$
   d. $2y + 4x = 1$

2. What is the range of the function $f(x) = x^2 + 9$?
   a. the set of all real numbers excluding 0
   b. the set of all real numbers excluding 3 and -3
   c. the set of all real numbers greater than or equal to 9
   d. the set of all real numbers greater than or equal to -9

3. Which of the following is the solution set for the inequality $-4 < 3x + 20 \leq 0$?
   a. $\left(-8, -\frac{20}{3}\right)$
   b. $\left[-8, -\frac{20}{3}\right] \cup \left\{\frac{20}{3}\right\}$
   c. $\left(-8, -\frac{20}{3}\right]$
   d. $\left[-8, -\frac{20}{3}\right)$

4. Which of the following are characteristics of the graph of

$f(x) = 1 - \frac{x^2 + 4}{x - 2}$?

   I. The function is equivalent to the linear function $g(x) = 1 - (x + 2)$ with a hole at $x = 2$.

   II. There is one vertical asymptote, no horizontal asymptote, and an oblique asymptote.

   III. There is one x-intercept and one y-intercept.

   a. I only

   b. II only

   c. II and III only

   d. I and III only

5. Which of the following, if any, are x-intercepts of the function

$f(x) = \ln\left(4x^2 - 4x + 2\right)$?

   a. $(\frac{1}{2}, 0)$

   b. $(-\frac{1}{2}, 0)$

   c. both **a** and **b**

   d. neither **a** nor **b**

6. Which of the following is the y-intercept of

$f(x) = -2\cos\left(2x + \frac{7\pi}{6}\right) - \frac{\sqrt{3}}{2}$?

   a. $(0, -2)$

   b. $\left(0, \frac{\sqrt{3}}{2}\right)$

   c. $\left(0, -\frac{\sqrt{3}}{2}\right)$

   d. $\left(-\frac{\sqrt{3}}{2}, 0\right)$

7. $\lim\limits_{x \to 2} \dfrac{-6x^2 - x + 1}{4x^3 + 4x^2 + x}$

   a. $\dfrac{1}{2}$

   b. $-\dfrac{1}{2}$

   c. 2

   d. The y-intercept does not exist.

8. On which of the following intervals is the graph of $f(x) = xe^{-2x}$ decreasing?

   a. $(-\infty, -1)$

   b. $\left(-\infty, \frac{1}{2}\right)$

   c. $(-1, \infty)$

   d. $\left(\frac{1}{2}, \infty\right)$

9. The height of a triangle increases by 4 feet every minute while its base shrinks by 7 feet every minute. How fast is the area changing when the height is 12 feet and the base is 18 feet?

10. Identify the locations of all local minima and maxima of $f(x) = \dfrac{3x^2}{x^2 + 2}$, where x can be any real number, using the first derivative test.

11. Which of the following is equivalent to $\dfrac{d}{dx} \displaystyle\int_{-4}^{x} \sqrt{t^2 + 3}\, dt$?

   a. $-\sqrt{x^2 + 3}$

   b. $\sqrt{x^2 + 3} - \sqrt{19}$

   c. $\sqrt{x^2 + 3}$

   d. $\sqrt{19} - \sqrt{x^2 + 3}$

12. If $f(x)$ is an even function, which of the following is equal to $\int_{-10}^{10} f^2(x)\,dx$?

 a. 0

 b. $2\int_0^{10} f^2(x)\,dx$

 c. both **a** and **b**

 d. neither **a** nor **b**

13. Compute $\int \dfrac{e^{5x}}{-4+e^{5x}}\,dx$.

14. The length of the portion of the curve $y = \cos x$ starting at $x = 0$ and ending at $x = \frac{\pi}{3}$ can be computed using which of the following?

 a. $\int_0^{\frac{\pi}{3}} \sqrt{1+(\cos x)^2}\,dx$

 b. $\int_0^{\frac{\pi}{3}} \sqrt{1+\cos x}\,dx$

 c. $\int_0^{\frac{\pi}{3}} \sqrt{1+\sin^2 x}\,dx$

 d. $\int_0^{\frac{\pi}{3}} \sqrt{1+\sin x}\,dx$

15. The family of curves described by the function $y(x) = C_1 e^{3x} + C_2 e^{-4x}$, where C_1 and C_2 are arbitrary real constants, satisfies which of the following differential equations?

 a. $y''(x) + y'(x) - 12y(x) = 0$

 b. $y''(x) - y'(x) - 12y(x) = 0$

 c. $\left(y'(x) - 3y(x)\right)\left(y'(x) + 4y(x)\right) = 0$

 d. $\left(y'(x) + 3y(x)\right)\left(y'(x) - 4y(x)\right) = 0$

16. If $a_n = 4 + (-1)^{n+1}$, $n \geq 1$, then which of the following is an accurate characterization of the series $\sum\limits_{n=1}^{\infty} a_n$?

a. The series $\sum\limits_{n=1}^{\infty} a_n$ converges with a sum of 3.

b. The series $\sum\limits_{n=1}^{\infty} a_n$ converges with a sum of 5.

c. The series $\sum\limits_{n=1}^{\infty} a_n$ diverges.

d. None of the above is true.

17. Which of the following is the interval of convergence for $\sum\limits_{n=0}^{\infty} \dfrac{x^{n+1}}{(5n+2)^2}$?
a. $[-1,1]$
b. $[-1,1)$
c. $(-1,1]$
d. $(-1,1)$

18. Which of the following Cartesian equations is equivalent to the polar equation $r^2 \cos 2\theta = 4$?
a. $y^2 - x^2 = 4$
b. $x^2 - y^2 = 4$
c. $2xy = 4$
d. $2x\sqrt{x^2 + y^2} = 4$

19. The graph of the parametrically defined curve, $x = -1 \sin^2 t$, $y = 2 + \cos^2 t$, $0 \leq t \leq \frac{\pi}{2}$ is a portion of a(n) _____.
a. circle
b. line
c. parabola
d. ellipse

20. True or false? Since the function $f(x) = |x - 5|$ is continuous at every real number x, it is also differentiable at every real number x.

Answers

1. d. The approach is to rewrite each equation in the slope-intercept form $y = mx + b$, where m is the slope of the line and b is the y-intercept. Doing so for choice **d** yields the equivalent equation $y = -2x + \frac{1}{2}$, which has a negative slope.

2. c. The domain of a real-valued function is the set of all values that, when substituted in for the variable, produces a meaningful output, while the range of a function is the set of all possible outputs. All real numbers can be substituted in for x in the function $f(x) = x^2 + 9$, so the domain of the function is the set of all real numbers. Since the x term is squared, the smallest value that this term can equal is 0 (when $x = 0$). Therefore, the smallest value that $f(x)$ can attain occurs when $x = 0$. Observe that $f(0) = 0^2 + 9 = 9$. Therefore, the range of $f(x)$ is the set of all real numbers greater than or equal to 9.

3. c. $-4 < 3x + 20 \leq 0$

$-24 < 3x \leq -20$

$-8 < x \leq -\frac{20}{3}$

So, the solution set is the half-open interval $\left(-8, -\frac{20}{3}\right]$.

4. b. First, note that $x^2 + 4$ does not factor, so $x = 2$ is a vertical asymptote for the graph of $f(x)$. This rules out I; $(x - 2)$ is not a factor of $x^2 + 4$, so $f(x)$ cannot be simplified to the equivalent of $g(x)$. Since the degree of the numerator of the fraction is exactly one more than that of the denominator, we can conclude that the graph has no horizontal asymptote, but does have an oblique asymptote. Hence, II is a characteristic of the graph of $f(x)$.

Next, while there is a y-intercept, (0,3), there is no x-intercept. To see this, we must consider the equation $f(x) = 0$, which is equivalent to $1 - \frac{x^2 + 4}{x - 2} = \frac{1(x-2) - (x^2 + 4)}{x - 2} = \frac{-(x^2 - x + 6)}{x - 2} = 0$ The x-values that satisfy such an equation are precisely those that make the numerator equal to zero *and* do not make the denominator equal to zero. Since the numerator does not factor, we know that $(x - 2)$ is not a factor of it; so we need to solve the equation $x^2 - x + 6 = 0$. Using the quadratic formula yields

$x = \frac{-(-1) \pm \sqrt{(-1)^2 - 4(1)(6)}}{2(1)} = \frac{1 \pm \sqrt{-23}}{2} = \frac{1 \pm i\sqrt{23}}{2}$

Since the solutions are imaginary, we conclude that there are no x-intercepts. Hence, III is not a characteristic of the graph of $f(x)$.

5. a. The x-intercepts of the function $f(x) = \ln\left(4x^2 - 4x + 2\right)$ are those x-values that satisfy the equation $4x^2 - 4x + 2 = 1$, solved as follows:

$$4x^2 - 4x + 2 = 1$$
$$4x^2 - 4x + 1 = 0$$
$$(2x - 1)^2 = 0$$

The solution of this equation is $x = \frac{1}{2}$, so the x-intercept is $(\frac{1}{2}, 0)$.

6. b. The y-intercept of $f(x) = -2\cos\left(2x + \frac{7\pi}{6}\right) - \frac{\sqrt{3}}{2}$ is the point $(0, f(0))$.

Observe that

$$f(0) = -2\cos\left(2(0) + \frac{7\pi}{6}\right) - \frac{\sqrt{3}}{2} = -2\cos\left(\frac{7\pi}{6}\right) - \frac{\sqrt{3}}{2} = -2\left(-\frac{\sqrt{3}}{2}\right) - \frac{\sqrt{3}}{2} = \frac{\sqrt{3}}{2}$$

Thus, the y-intercept is $\left(0, \frac{\sqrt{3}}{2}\right)$.

7. b. To compute this limit, factor the numerator and denominator, cancel factors that are common to both, and substitute $x = 2$ into the simplified expression, as follows:

$$\lim_{x \to 2} \frac{-6x^2 - x + 1}{4x^3 + 4x^2 + x} = \lim_{x \to 2} \frac{(1 - 3x)(2x + 1)}{x(2x + 1)(2x + 1)} = \lim_{x \to 2} \frac{1 - 3x}{x(2x + 1)}$$

$$= \frac{1 - 3(2)}{2(2(2) + 1)} = -\frac{5}{10} = -\frac{1}{2}$$

8. d. The graph of $f(x) = xe^{-2x}$ is decreasing on the interval where $f'(x) < 0$. We compute the derivative of $f(x)$ as follows:

$$f'(x) = x \cdot \left(-2e^{-2x}\right) + (1)\left(e^{-2x}\right) = e^{-2x}(-2x + 1).$$

Since e^{-2x} is always positive, the only x-values for which $f'(x) < 0$ are those for which $-2x + 1 < 0$, or $x > \frac{1}{2}$; that is, for only those x-values in the interval $\left(\frac{1}{2}, \infty\right)$.

9. The formula for the area of a triangle with base b and height h is $A = \frac{1}{2}bh$. Implicitly differentiating both sides with respect to time t yields

$$\frac{dA}{dt} = \frac{1}{2} \cdot \frac{db}{dt} \cdot h + \frac{dh}{dt} \cdot \frac{1}{2}b$$

Suppressing units, we are given $\frac{dh}{dt} = 4$ and $\frac{db}{dt} = -7$, and are asked to determine $\frac{dA}{dt}$ at the instant in time when $h = 12$ and $b = 18$.

Substituting these values into the expression for $\frac{dA}{dt}$ yields $\frac{dA}{dt} = -6$, meaning that the area is decreasing at the rate of 6 square feet per minute at this particular instant.

10. Applying the first derivative test requires that we compute the first derivative, as follows:

$$f'(x) = \frac{(x^2 + 2)(6x) - 3x^2(2x)}{(x^2 + 2)^2} = \frac{6x^3 + 12x - 6x^3}{\left(x^2 + 2\right)^2} = \frac{12x}{\left(x^2 + 2\right)^2}$$

Observe that there are no x-values at which $f'(x)$ is undefined, and the only x-value that makes $f'(x) = 0$ is $x = 0$. To assess how the sign of $f'(x)$ changes, we form a number line using the critical points, choose a value in each subinterval duly formed, and report the sign of $f'(x)$ above the subinterval, as follows:

$$f'(x) \quad - \quad | \quad + $$
$$0$$

Since the sign of $f'(x)$ changes from $-$ to $+$ at $x = 0$, we conclude that $f(x)$ has a local minimum at this value. There is no local maximum.

11. c. Using the fundamental theorem of calculus immediately yields

$$\frac{d}{dx} \int_{-4}^{x} \sqrt{t^2 + 3} \; dt = \sqrt{x^2 + 3}$$

12. b. Since $f(x)$ is an even function, then $f(x) = f(-x)$. Therefore, $f^2(x) = f^2(-x)$, so $f^2(x)$ is also an even function. Hence, the symmetry property of the integral implies that

$$\int_{-10}^{10} f^2(x)dx = 2\int_{0}^{10} f^2(x)dx$$

13. Make the following substitution:

$$u = -4 + e^{5x}$$

$$du = 5e^{5x}\, dx \implies \tfrac{1}{5} du = e^{5x}\, dx$$

Applying this substitution in the integrand and computing the resulting indefinite integral yields

$$\int \frac{e^{5x}}{-4+e^{5x}}\, dx = \int \frac{1}{-4+e^{5x}}\left(e^{5x}\right) dx = \int \frac{1}{u}\left(\frac{1}{5}\right) du = \frac{1}{5}\int \frac{1}{u} du = \frac{1}{5}\ln|u| + C$$

Finally, rewrite the final expression in terms of the original variable x by resubstituting $u = -4 + e^{5x}$ to obtain

$$\int \frac{e^{5x}}{-4+e^{5x}}\, dx = \frac{1}{5}\ln\left|-4+e^{5x}\right| + C$$

14. c. Since $\frac{dy}{dx} = -\sin x$, we apply the length formula $\int_a^b \sqrt{1+\left(f'(x)\right)^2}\, dx$ to conclude that the length of the portion of the curve $y = \cos x$ starting at $x = 0$ and ending at $x = \frac{\pi}{3}$ can be computed using the integral $\int_0^{\frac{\pi}{3}} \sqrt{1+\sin^2 x}\, dx$.

15. a. Solving this question requires the process of elimination. Observe that $y'(x) = 3C_1 e^{3x} - 4C_2 e^{-4x}$ and $y''(x) = 9C_1 e^{3x} + 16C_2 e^{-4x}$. Substituting these functions, along with the given function $y(x) = C_1 e^{3x} + C_2 e^{-4x}$, into the differential equation provided in choice **a** yields

$$y''(x) + y'(x) - 12y(x) = \left(9C_1 e^{3x} + 16C_2 e^{-4x}\right) + \left(3C_1 e^{3x} - 4C_2 e^{-4x}\right)$$

$$-12\left(C_1 e^{3x} + C_2 e^{-4x}\right)$$

$$= \left(9C_1 + 3C_1 - 12C_1\right)e^{3x}$$

$$+ \left(16C_2 - 4C_2 - 12C_2\right)e^{-4x}$$

$$= 0$$

16. c. Note that the terms of the sequence $\{a_n\}$ can be expressed by the list $5, 3, 5, 3, \ldots$ As such, the sequence does not converge. Specifically, $\lim_{n\to\infty} a_n \neq 0$. As a result, we conclude from the nth term test for divergence that the series $\sum_{n=1}^{\infty} a_n$ diverges.

17. a. The interval of convergence is determined by applying the ratio test, as follows:

$$\lim_{n\to\infty}\left|\frac{\frac{x^{n+2}}{(5(n+1)+2)^2}}{\frac{x^{n+1}}{(5n+2)^2}}\right| = \lim_{n\to\infty}\left|\frac{x^{n+2}}{(5(n+1)+2)^2}\cdot\frac{(5n+2)^2}{x^{n+1}}\right| = |x|\lim_{n\to\infty}\left|\frac{(5n+2)^2}{(5n+7)^2}\right| = |x|$$

Now, we determine the values of x that make the result of this test < 1. This is given by the inequality $|x| < 1$, or equivalently $-1 < x < 1$. Therefore, the power series converges at least for all x in the interval $(-1,1)$.

The endpoints must be checked separately to determine if they should be included in the interval of convergence of the power series $\sum_{n=0}^{\infty}\frac{x^{n+1}}{(5n+2)^2}$. First, substituting $x = 1$ results in the series $\sum_{n=0}^{\infty}\frac{1}{(5n+2)^2}$.

Using the limit comparison test with the convergent p-series $\sum_{n=1}^{\infty}\frac{1}{n^2}$,

we conclude that since $\lim_{n\to\infty}\frac{\frac{1}{(5n+2)^2}}{\frac{1}{n^2}} = \lim_{n\to\infty}\frac{n^2}{(5n+2)^2} = \lim_{n\to\infty}\frac{n^2}{25n^2+20n+4}$

$= \frac{1}{25} > 0$, it follows that $\sum_{n=0}^{\infty}\frac{1}{(5n+2)^2}$ converges. So, $x = 1$ is included in the interval of convergence. As for $x = -1$, we must determine if the series $\sum_{n=0}^{\infty}\frac{(-1)^{n+1}}{(5n+2)^2}$ converges. Applying the alternating series test, note that since the sequence $\left\{\frac{1}{(5n+2)^2}\right\}_{n=0}^{\infty}$ consists of positive terms that decrease toward zero, we conclude that $\sum_{n=0}^{\infty}\frac{(-1)^{n+1}}{(5n+2)^2}$ converges. So, $x = -1$ is included in the interval of convergence.

Hence, we conclude that the interval of convergence is $[-1,1]$.

18. b. Applying the double-angle formula $\cos 2\theta = \cos^2\theta - \sin^2\theta$ into the original equation, simplifying, and applying the transformation equations $x = r\cos\theta$ and $y = r\sin\theta$ yields:

$r^2\cos 2\theta = 4$

$r^2\cos^2\theta - r^2\sin^2\theta = 4$

$x^2 - y^2 = 4$

19. b. The given parametric equations are equivalent to $x + 1 = \sin^2 t$ and $y - 2 = \cos^2 t$, $0 \le t \le \frac{\pi}{2}$. We can eliminate the parameter t by using the trigonometric identity $\sin^2 \theta + \cos^2 \theta = 1$. Doing so yields the equivalent Cartesian equation $(x + 1) + (y - 2) = 1$, or, equivalently, $y = -x + 2$; this is a line. The given curve is the portion of this line starting at the point $(-1, 3)$ (when $t = 0$) and ending at the point $(0, 2)$ (when $t = \frac{\pi}{2}$).

20. False. Continuity does not imply differentiability. The function $f(x) = |x - 5|$ is continuous at every real number x, but it is not differentiable at $x = 5$ because its graph has a sharp corner at that point.

Algebra Review Problems

Understanding the basic rules of algebra and developing a comfortable working knowledge of them are crucial to the study of calculus. Among the most common algebraic tools are simplifying algebraic expressions involving exponents, performing basic arithmetic operations on polynomial expressions, simplifying radical expressions, and solving a wide variety of equations. These techniques are reviewed in this section.

Questions

1. What is the value of the expression $-4(3a^2 - 24)$ when $a = -3$?

2. Simplify: $4(k + m) - 9(3k - 2m) + k + m$

3. Simplify: $-3\{2a(-4a + 3) + 6(4 - a)\} + 5a^2$

4. Simplify: $(9a^6 b^{11} c^4)(9a^2 c^2)$

5. Simplify: $\dfrac{-42m^2 n^{-2} p^5}{6mn^2}$

6. Simplify: $(3x^{-\frac{3}{2}} y^5)^2 + (\frac{1}{2}x)^{-3} y^2 (-2y^4)^2$

7. Simplify: $\dfrac{2(3x^{-2} y)^{-2} (x^{-1} y)^3}{-(x^{-2} y^3)^{-2}}$

8. Simplify: $(-2x^2 y)^3 - \dfrac{(2x^2 y)^4}{2x^2 y}$

9. Multiply: $2x(5x^2 - 6x - 3)$

10. Multiply: $(3x - 2)(4x + 1)$

11. Multiply: $(x + 2)(3x^2 - 5x + 2)$

12. Multiply: $(3y - 4)(y + 3)(5y + 2)$

13. Multiply: $(2x^2 + 5)(x - 2)(2x^2 - 5)$

14. Factor completely: $15y - 60y^3$

15. Factor completely: $x^2 + 7x + 12$

16. Factor completely: $v^4 - 13v^2 - 48$

17. Factor completely: $7x^4 y^2 - 35x^3 y^3 + 42x^2 y^4$

18. Factor completely: $4x^{-8} + 8x^{-4} + 3$

19. Simplify: $-4\sqrt[3]{27k^6}$

20. Simplify: $\sqrt[4]{32w^3x^7y^{-6}z^{15}}$

21. Solve for x: $3\sqrt{x} = 18$

22. Solve for d: $\sqrt[3]{5d-4} = 5$

23. Solve for z: $z = \sqrt{z+12}$

24. Solve by factoring: $49r^2 = 100$

25. Solve by factoring: $x^2 + 12x + 32 - 0$

26. Solve by factoring: $v^6 - 13v^3 - 48 = 0$

27. Solve by using the quadratic formula: $18x^2 + 9x + 1 = 0$

28. Solve by using the quadratic formula: $x^2 + 10x + 11 = 0$

29. Solve by using radical methods: $(x+4)^2 - 9 = 0$

30. Solve by using radical methods: $(5x - 3)^2 - 4 = 8$

31. Solve by completing the square: $x^2 + 12x + 32 = 0$

32. Solve by completing the square: $18x^2 + 3x + 1 = 0$

33. Solve for r: $V = 4\pi r^2 h$, assuming that $r > 0$

34. Solve for w: $P = 2l + 2w$

35. Solve for z: $az + bx^2 = a(d - z)$, assuming that $a > 0$

Answers

1. Apply the usual order of operations as follows:
$$-4(3a^2 - 24) = -4[3(-3)^2 - 24] = -4[3(9) - 24] = -4[27 - 24] = -4[3]$$
$$= -12$$

2. Apply the distributive law of multiplication and gather like terms (i.e., those terms with the same variable) as follows:
$$4(k + m) - 9(3k - 2m) + k + m = 4k + 4m - 27k + 18m + k + m$$
$$= (4 - 27 + 1)k + (4 + 18 + 1)m$$
$$= -22k + 23m$$

3. Apply the distributive law of multiplication, together with the usual order of operations, and gather like terms as follows:
$$-3\{2a(-4a + 3) + 6(4 - a)\} + 5a^2 = -3\{-8a^2 + 6a + 24 - 6a\} + 5a^2$$
$$= -3\{-8a^2 + 24\} + 5a^2$$
$$= 24a^2 - 72 + 5a^2 = 29a^2 - 72$$

4. Since multiplication is commutative, we can group together terms with the same base. Doing so, and applying the exponent rule for multiplying terms with like bases, yields the following:
$$\left(9a^6 b^{11} c^4\right)\left(9a^2 c^2\right) = (9 \cdot 9)\left(a^6 \cdot a^2\right)\left(b^{11}\right)\left(c^4 \cdot c^2\right) = 81a^8 b^{11} c^6$$

5. Since multiplication is commutative, we can group together terms with the same base. Doing so, and applying the exponent rule for dividing terms with like bases, yields the following:
$$\frac{-42m^2 n^{-2} p^5}{6mn^2} = \left(\frac{-42}{6}\right)\left(\frac{m^2}{m}\right)\left(\frac{n^{-2}}{n^2}\right)\left(p^5\right) = -7(m)\left(\frac{1}{n^4}\right)\left(p^5\right) = \frac{-7mp^5}{n^4}$$

6. Apply the exponent rules, together with the order of operations, as follows:
$$\left(3x^{-\frac{3}{2}} y^5\right)^2 + \left(\tfrac{1}{2}x\right)^{-3} y^2 \left(-2y^4\right)^2 = 3^2\left(x^{-\frac{3}{2}}\right)^2 \left(y^5\right)^2 + \left(\tfrac{1}{2}\right)^{-3} x^{-3} y^2 (-2)^2 \left(y^4\right)^2$$
$$= 9x^{-3} y^{10} + 8x^{-3} y^2 (4) y^8$$
$$= 9x^{-3} y^{10} + 32x^{-3} y^{10} = 41x^{-3} y^{10}$$

7. Apply the exponent rules, together with the order of operations, as follows:
$$\frac{2\left(3x^{-2} y\right)^{-2}\left(x^{-1} y\right)^3}{-\left(x^{-2} y^3\right)^{-2}} = \frac{2(3)^{-2}\left(x^{-2}\right)^{-2} y^{-2}\left(x^{-1}\right)^3 y^3}{-\left(x^{-2}\right)^{-2}\left(y^3\right)^{-2}} = \frac{2\left(\tfrac{1}{9}\right)x^4 y^{-2} x^{-3} y^3}{-x^4 y^{-6}} = \frac{\tfrac{2}{9}xy}{-x^4 y^{-6}} = -\frac{2y^7}{9x^3}$$

8. Although you can apply the exponent rules as in the previous two questions, it is more efficient to treat the expression $2x^2y$ as a single term as long as possible. Specifically, we have:

$$\left(-2x^2y\right)^3 - \frac{\left(2x^2y\right)^4}{2x^2y} = (-1)^3\left(2x^2y\right)^3 - \frac{\left(2x^2y\right)^4}{\left(2x^2y\right)} = -\left(2x^2y\right)^3 - \left(2x^2y\right)^3$$

$$= -2\left(2x^2y\right)^3 = -2\left(2^3\right)\left(x^2\right)^3 y^3 = -2(8)x^6y^3$$

$$= -16x^6y^3$$

9. Applying the distributive law of multiplication, together with the exponent rule for multiplying terms with the same base, yields

$$2x\left(5x^2 - 6x - 3\right) = 2x\left(5x^2\right) + 2x(-6x) + 2x(-3) = 10x^3 - 12x^2 - 6x$$

10. Use the distributive property first to multiply the binomial $(4x+1)$, as a single expression, by each term of $(3x-2)$. Then, apply the distributive law to multiply each term of $(4x+1)$ by $3x$ and -2. Finally, combine like terms. Doing so yields

$$(3x-2)(4x+1) = 3x(4x+1) - 2(4x+1)$$

$$= 3x(4x) + 3x(1) - 2(4x) - 2(1)$$

$$= 12x^2 + 3x - 8x - 2 = 12x^2 - 5x - 2$$

11. Use the distributive property first to multiply the trinomial $\left(3x^2 - 5x + 2\right)$, as a single expression, by each term of $(x+2)$. Then, apply the distributive law to multiply each term of $\left(3x^2 - 5x + 2\right)$ by x and 2. Finally, combine like terms. Doing so yields

$$(x+2)\left(3x^2 - 5x + 2\right) = x\left(3x^2 - 5x + 2\right) + 2\left(3x^2 - 5x + 2\right)$$

$$= x\left(3x^2\right) + x(-5x) + x(2) + 2\left(3x^2\right) + 2(-5x) + 2(2)$$

$$= 3x^3 - 5x^2 + 2x + 6x^2 - 10x + 4$$

$$= 3x^3 + x^2 - 8x + 4$$

12. First, multiply the first two binomials in the expression using the distributive law as in Question 10. Doing so results in a trinomial, as follows:

$$(3y-4)(y+3)=3y(y+3)-4(y+3)=3y(y)+3y(3)-4(y)-4(3)$$
$$=3y^2+9y-4y-12=3y^2+5y-12$$

Now, multiply the trinomial $(3y^2+5y-12)$ by the binomial $(5y+2)$ as in Question 11, as follows:

$$(3y-4)(y+3)(5y+2)=(3y^2+5y-12)(5y+2)$$
$$=(5y+2)(3y^2+5y-12)$$
$$=5y(3y^2+5y-12)+2(3y^2+5y-12)$$
$$=5y(3y^2)+5y(5y)+5y(-12)+2(3y^2)$$
$$\quad+2(5y)+2(-12)$$
$$=15y^3+25y^2-60y+6y^2+10y-24$$
$$=15y^3+31y^2-50y-24$$

13. While you could proceed in the exact same manner as in Question 12, it is beneficial to first switch the order of the product to obtain the equivalent expression $(2x^2+5)(2x^2-5)(x-2)$. This is advantageous because the product of the first two binomials is of the form $(a+b)(a-b)$, which upon simplification is equal to the binomial a^2-b^2. Had we not switched the order of the terms in the original expression, the product of the first two binomials would have been a trinomial, thereby creating more work in the second stage of the multiplication process. Using this approach, the multiplication is as follows:

$$(2x^2+5)(x-2)(2x^2-5)=(2x^2+5)(2x^2-5)(x-2)$$
$$=\left[(2x^2)^2-5^2\right](x-2)=(4x^4-25)(x-2)$$
$$=4x^4(x-2)-25(x-2)$$
$$=4x^4(x)-4x^4(2)-25(x)+25(2)$$
$$=4x^5-8x^4-25x+50$$

14. First, factor out the greatest common factor (GCF), $15y$. Then, factor the resulting binomial using the difference of squares formula $(a+b)(a-b)=a^2-b^2$, as follows:

$$15y-60y^3=15y\left(1-4y^2\right)=15y\left(1-(2y)^2\right)=15y\left(1-2y\right)\left(1+2y\right)$$

15. This expression can be factored using the trinomial method. Since the factors of x^2 are simply x and x, the goal is to find integers c and d such that

$$x^2+7x+12=(x+c)(x+d)$$

This is satisfied only when both $c+d=7$ and $c\cdot d=12$ hold simultaneously. Observe that this holds when $c=3$ and $d=4$. Hence, we conclude that

$$x^2+7x+12=(x+3)(x+4)$$

16. This expression can be factored using the trinomial method. Two factors of v^4 are v^2 and v^2. We attempt to find integers c and d such that

$$v^4-13v^2-48=\left(v^2+c\right)\left(v^2+d\right)$$

This is satisfied only when both $c+d=-13$ and $c\cdot d=48$ hold simultaneously. There are several pairs of factors to consider, but the only pair that works is when $c=3$ and $d=-16$. Hence, we have

$$v^4-13v^2-48=\left(v^2+3\right)\left(v^2-16\right)$$

We are not yet finished, however, since the binomial $\left(v^2-16\right)$ is of the form $a^2-b^2=(a+b)(a-b)$, and hence factors as $(v-4)(v+4)$. The first binomial $\left(v^2+3\right)$ does not factor into a product of linear terms involving only real constants. Therefore, we conclude that

$$v^4-13v^2-48=\left(v^2+3\right)(v-4)(v+4)$$

17. First, factor out the GCF to obtain

$$7x^4y^2 - 35x^3y^3 + 42x^2y^4 = 7x^2y^2\left(x^2 - 5xy + 6y^2\right)$$

Next, we apply the trinomial method to factor $x^2 - 5xy + 6y^2$. Since the factors of x^2 are simply x and x, the goal is to find integers c and d such that

$$x^2 - 5xy + 6y^2 = (x + cy)(x + dy)$$

This is satisfied only when both $c + d = -5$ and $c \cdot d = 6$ hold simultaneously. Observe that this holds when $c = -3$ and $d = -2$. Hence, we conclude that

$$x^2 - 5xy + 6y^2 = (x - 3y)(x - 2y)$$

Thus, the complete factorization of the original expression is given by

$$7x^4y^2 - 35x^3y^3 + 42x^2y^4 = 7x^2y^2(x - 3y)(x - 2y)$$

18. First, make the substitution $u = x^{-4}$ to rewrite the given expression as follows:

$$4x^{-8} + 8x^{-4} + 3 = 4\left(x^{-4}\right)^2 + 8\left(x^{-4}\right) + 3 = 4u^2 + 8u + 3$$

The idea is to now factor the expression on the right side, and then resubstitute $u = x^{-4}$ into the factored expression.

We apply the trinomial method to factor $4u^2 + 8u + 3$. Two factors of $4u^2$ are $2u$ and $2u$. We attempt to find integers c and d such that

$$4u^2 + 8u + 3 = (2u + c)(2u + d)$$

This is satisfied only when both $2c + 2d = 8$ and $c \cdot d = 3$ hold simultaneously. Observe that this holds when $c = 3$ and $d = 1$. Hence, we conclude that

$$4u^2 + 8u + 3 = (2u + 3)(2u + 1)$$

Now, resubstitute $u = x^{-4}$ into the factored expression to obtain

$$4x^{-8} + 8x^{-4} + 3 = \left(2x^{-4} + 3\right)\left(2x^{-4} + 1\right)$$

Neither of the two binomials on the right side factors further over the set of real numbers.

19. First, rewrite the expression underneath the radical sign as a product of terms raised to the third power and terms whose powers are less than 3 to obtain

$$-4\sqrt[3]{27k^6} = -4\sqrt[3]{3^3\left(k^2\right)^3}$$

Now, simplify the radical expression using the fact that $\sqrt[3]{a^3} = a$ to obtain the simplified expression

$$-4\sqrt[3]{27k^6} = -4\left(3k^2\right) = -12k^2$$

20. First, rewrite the expression underneath the radical sign as a product of terms raised to the fourth power and terms whose powers are less than 4 to obtain

$$\sqrt[4]{32w^3x^7y^{-6}z^{15}} = \sqrt[4]{2^4 \cdot 2w^3x^4x^3\left(y^{-1}\right)^4 y^{-2}\left(z^3\right)^4 z^3}$$
$$= \sqrt[4]{\left(2xy^{-1}z^3\right)^4 \cdot 2w^3x^3y^{-2}z^3}$$

Now, simplify the radical expression using the fact that $\sqrt[4]{a^4} = a$ to obtain the simplified expression

$$\sqrt[4]{\left(2xy^{-1}z^3\right)^4 \cdot 2w^3x^3y^{-2}z^3} = \left(2xy^{-1}z^3\right)\sqrt[4]{2w^3x^3y^{-2}z^3}$$

21. Isolate the radical term on one side of the equation by dividing by 3, then square both sides to obtain

$$3\sqrt{x} = 18$$
$$\sqrt{x} = 6$$
$$x = 36$$

22. Since the radical term is already isolated, we simply cube both sides and then solve for d, as follows:

$$\sqrt[3]{5d-4} = 5$$
$$5d-4 = 125$$
$$5d = 129$$
$$d = \frac{129}{5}$$

23. Since the radical term is already isolated, we begin by squaring both sides, and subsequently moving all terms to the left side (so that the coefficient of the term with the largest exponent is positive).

$z = \sqrt{z+12}$

$z^2 = z+12$

$z^2 - z - 12 = 0$

Now, factor the expression on the left using the trinomial method to obtain the equivalent equation

$(z-4)(z+3) = 0$

Since the product of two real numbers is zero if and only if one of the numbers itself is zero, we conclude that at least one of $(z-4)$ and $(z+3)$ is zero, so that the solutions of the factored equation are $z = -3$ and $z = 4$. However, we must finally determine which of these satisfies the original equation since squaring both sides of equations and subsequently solving *that* equation could produce z-values that don't satisfy the original equation. Upon substituting each of them into the original equation, we find that $z = 4$ results in a true statement, and so is a solution, whereas $z = -3$ results in the false statement $-3 = \sqrt{9}$, and so is not a solution. Therefore, we conclude that the only solution to the original equation is $z = 4$.

24. First, move all terms to the left side. Then, use the difference of squares formula $(a+b)(a-b) = a^2 - b^2$ to factor the expression on the left side, as follows:

$49r^2 = 100$

$49r^2 - 100 = 0$

$(7r)^2 - 10^2 = 0$

$(7r-10)(7r+10) = 0$

Since the product of two real numbers is zero if and only if one of the numbers itself is zero, we conclude that at least one of $(7r-10)$ and $(7r+10)$ is zero, so that the solutions of the factored equation are $r = \pm\frac{10}{7}$. Substituting each of these into the original equation results in a true statement, so they are both solutions of the original equation.

25. The expression on the left side of the equation can be factored using the trinomial method. Since the factors of x^2 are simply x and x, the goal is to find integers c and d such that

$x^2 + 12x + 32 = (x + c)(x + d)$

This is satisfied only when both $c + d = 12$ and $c \cdot d = 32$ hold simultaneously. Observe that this holds when $c = 4$ and $d = 8$. Hence, we conclude that the original equation is equivalent to

$(x + 4)(x + 8) = 0$

Since the product of two real numbers is zero if and only if one of the numbers itself is zero, we conclude that at least one of $(x + 4)$ and $(x + 8)$ is zero, so that the solutions of the factored equation are $x = -4$ and $x = -8$. Substituting each of these into the original equation results in a true statement, so they are both solutions of the original equation.

26. The expression on the left side of the equation can be factored using the trinomial method. Two factors of v^6 are v^3 and v^3. We attempt to find integers c and d such that

$v^6 - 13v^3 - 48 = \left(v^3 + c\right)\left(v^3 + d\right)$

This is satisfied only when both $c + d = -13$ and $c \cdot d = -48$ hold simultaneously. There are several pairs of factors to consider, but the only pair that works is when $c = 3$ and $d = -16$. Hence, we have

$v^6 - 13v^3 - 48 = \left(v^3 + 3\right)\left(v^3 - 16\right)$

The original equation is equivalent to $\left(v^3 + 3\right)\left(v^3 - 16\right) = 0$. Since the product of two real numbers is zero if and only if one of the numbers itself is zero, we conclude that at least one of $v^3 + 3 = 0$ and $v^3 - 16 = 0$ holds. We solve each of these equations by moving the constant term to the right side and then taking the cube root, thus concluding that the solutions to the equation we seek are $v = \sqrt[3]{-3}$ and $v = \sqrt[3]{16} = 2\sqrt[3]{2}$. Substituting each of these into the original equation results in a true statement, so they are both solutions of the original equation.

27. The solutions of the quadratic equation $ax^2 + bx + c = 0$, where $a \neq 0$,

are given by the quadratic formula $x = \frac{-b \pm \sqrt{b^2 - 4ac}}{2a}$.

Applying this formula with $a = 18$, $b = 9$, $c = 1$ yields

$x = \frac{-9 \pm \sqrt{9^2 - 4(18)(1)}}{2(18)} = \frac{-9 \pm \sqrt{81-72}}{36} = \frac{-9 \pm \sqrt{9}}{36} = \frac{-9 \pm 3}{36}$

Simplifying the right side yields

$x = \frac{-9-3}{36} = -\frac{12}{36} = -\frac{1}{3}, \quad x = \frac{-9+3}{36} = -\frac{6}{36} = -\frac{1}{6}$

Both of these values are solutions to the original equation.

28. The solutions of the quadratic equation $ax^2 + bx + c = 0$, where $a \neq 0$,

are given by the quadratic formula $x = \frac{-b \pm \sqrt{b^2 - 4ac}}{2a}$.

Applying this formula with $a = 1$, $b = 10$, $c = 11$ yields

$x = \frac{-10 \pm \sqrt{10^2 - 4(1)(11)}}{2(1)} = \frac{-10 \pm \sqrt{100 - 44}}{2} = \frac{-10 \pm \sqrt{56}}{2} = \frac{-10 \pm 2\sqrt{14}}{2}$

Simplifying the right side yields

$x = \frac{-10 - 2\sqrt{14}}{2} = -5 - \sqrt{14}, \quad x = \frac{-10 + 2\sqrt{14}}{2} = -5 + \sqrt{14}$

Both of these values are solutions to the original equation.

29. First, take the constant term, 9, to the right side. Then, take the square root of both sides and solve for x, as follows:

$(x+4)^2 - 9 = 0$

$(x+4)^2 = 9$

$x + 4 = \pm\sqrt{9} = \pm 3$

$x = -4 \pm 3$

Simplifying the right side yields

$x = -4 - 3 = -7, \quad x = -4 + 3 = -1$

Both of these values are solutions of the original equation.

30. First, add 4 to both sides of the equation. Then, take the square root of both sides and solve for x, as follows:

$(5x-3)^2 - 4 = 8$

$(5x-3)^2 = 12$

$5x - 3 = \pm\sqrt{12} = \pm 2\sqrt{3}$

$5x = 3 \pm 2\sqrt{3}$

$x = \frac{3 \pm 2\sqrt{3}}{5}$

Both of these values are solutions to the original equation.

31. The strategy behind completing the square is to rewrite a quadratic equation of the form $ax^2 + bx + c = 0$ as an equivalent equation of the form $a(x+e)^2 + f = 0$, where a, b, c, e, and f are real numbers and $a \neq 0$. The reason is that the latter equation is easily solved using radical methods. In order to make this transformation, we must determine which real number w can be added to the expression $x^2 + 12x$ so that $x^2 + 12x + w$ factors as a perfect square $(x+e)^2$. This value w is chosen to be $\left(\frac{1}{2} \cdot 12\right)^2$, or 36. Indeed, observe that $x^2 + 12x + 36 = (x+6)^2$. Now, so as to not change the value of the left side, we must add *and* subtract 36; then, we proceed to solve the resulting equation, as follows:

$x^2 + 12x + 32 = 0$

$\left(x^2 + 12x + 36\right) - 36 + 32 = 0$

$(x+6)^2 - 4 = 0$

$(x+6)^2 = 4$

$x + 6 = \pm\sqrt{4} = \pm 2$

$x = -6 \pm 2$

Hence, the solutions of the original equation are
$x = -6 - 2 = -8, \quad x = -6 + 2 = -4$

32. The strategy behind completing the square is to rewrite a quadratic equation of the form $ax^2 + bx + c = 0$ as an equivalent equation of the form $a(x+e)^2 + f = 0$, where a, b, c, e, and f are real numbers and $a \neq 0$. The reason is that the latter equation is easily solved using radical methods. In order to make this transformation, we first factor the coefficient of x^2, namely 18, out of the first two terms only to obtain:

$18x^2 + 3x + 1 = 0$

$18\left(x^2 + \frac{1}{6}x\right) + 1 = 0$

Next, we must determine which real number w can be added to the expression $x^2 + \frac{1}{6}x$ so that $x^2 + \frac{1}{6}x + w$ factors as a perfect square $(x+e)^2$. This value w is chosen to be $\left(\frac{1}{2} \cdot \frac{1}{6}\right)^2$, or $\frac{1}{144}$. Indeed, observe that $x^2 + \frac{1}{6}x + \frac{1}{144} = \left(x + \frac{1}{12}\right)^2$. Now, as to not change the value of the left side, we must add *and* subtract the same constant. But, keep in mind that we added $\frac{1}{144}$ to an expression in parentheses multiplied by 18. Hence, we actually need to add the constant $18\left(\frac{1}{144}\right)$ to and subtract it from the left side. Then, we proceed to solve the resulting equation, as follows:

$18x^2 + 3x + 1 = 0$

$18\left(x^2 + \frac{1}{6}x\right) + 1 = 0$

$18\left(x^2 + \frac{1}{6}x + \frac{1}{144}\right) - 18\left(\frac{1}{144}\right) + 1 = 0$

$18\left(x + \frac{1}{12}\right)^2 + \frac{126}{144} = 0$

$18\left(x + \frac{1}{12}\right)^2 = -\frac{126}{144}$

Note that the left side of the equation is nonnegative, but the right side is negative. Therefore, there can be no value of x that satisfies the equation. Hence, there is no solution.

33. Since the right side is simply a product of terms, we begin by dividing both sides by all terms except r^2. Then, we take the square root, as follows.

$V = 4\pi r^2 h$

$\frac{V}{4\pi h} = r^2$

$\sqrt{\frac{V}{4\pi h}} = r$

Note that taking the square root of both sides technically produces two results, namely the expression and its negative. However, since the restriction that $r > 0$ is imposed, we discard the negative expression.

34. Isolate *w* on one side of the equation, as follows:

$$P = 2l + 2w$$
$$P - 2l = 2w$$
$$\frac{P - 2l}{2} = w$$

35. Gather all terms involving *z* on one side of the equation and all other terms on the other side. Then, solve for *z* as follows:

$$az + bx^2 = a(d - z)$$
$$az + bx^2 = ad - az$$
$$az + az = ad - bx^2$$
$$2az = ad - bx^2$$
$$z = \frac{ad - bx^2}{2a}$$

Linear Equations and Inequality Problems

Linear equations and their graphs play a critical role in the study of calculus. The strategy employed when solving linear equations simply involves simplifying various expressions involved by clearing fractions and using the order of operations, together with the distributive property of multiplication, to isolate the variable on one side of the equation. The same basic strategy is used to solve linear inequalities, with the one additional feature that the inequality sign is switched whenever both sides of the inequality are multiplied by a negative real number.

Graphing a line requires the slope m of the line and a point (x_1, y_1) that is known to be on the line. This information can be used to write the equation of the line in the *point-slope form* $y - y_1 = m(x - x_1)$, and solving this equation for y and simplifying yields the *slope-intercept* form of the line, $y = mx + b$, where b is the y-intercept. A line with a positive slope rises from left to right, whereas one with a negative slope falls from left to right. The slope of a horizontal line is zero, whereas a vertical line does not have a defined slope.

Sometimes, linear equations and inequalities also involve terms in which the absolute value is taken of a linear expression. The *absolute value* of a is defined by

$$|a| = \begin{cases} a, \text{ if } a \geq 0 \\ -a, \text{ if } a < 0 \end{cases}$$

Note that the absolute value of a quantity is necessarily nonnegative.

Properties of Absolute Value

1. $|a| = b$ if and only if $a = \pm b$.
2. $|a| = |b|$ if and only if $a = \pm b$.
3. $|a| \geq c$ if and only if $a \geq c$ or $a \leq -c$.
4. $|a| < c$ if and only if $-c < a < c$.

Questions

36. Solve for x: $10x - 2(3 - x) = -3(4 + 2x) - 2x$

37. Solve for x: $8\left(\frac{1}{4}x - \frac{1}{2}\right) = -2(2 - \frac{1}{4}x)$

38. Solve for x: $0.8(x + 20) - 4.5 = 0.7(5 + x) - 0.9x$

39. Graph the line whose equation is given by $y = 2x + 3$.

40. Graph the line whose equation is given by $x - 3y = 12$.

41. Graph the line whose equation is given by $-\frac{5}{3}x - \frac{1}{3}y = -2$.

42. Choose the equation that represents the graph in the following figure.

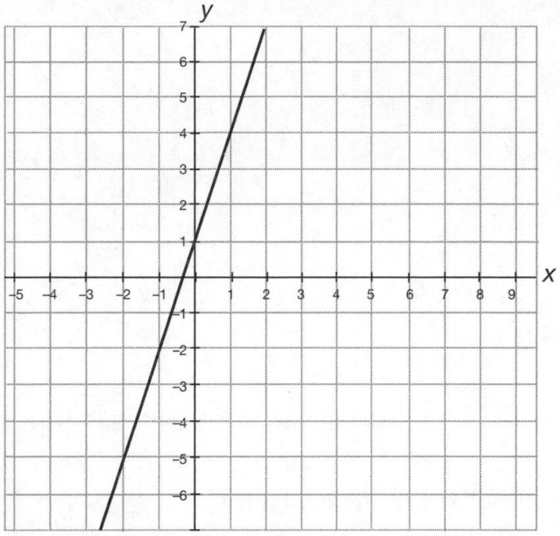

    a. $y = 3x + 1$
    b. $y = -3x - 1$
    c. $y = 3x - 1$
    d. $y = \frac{1}{3}x + 1$

43. Which of the following linear equations has a negative slope?
    a. $y = 3x - 5$
    b. $-1 = y - x$
    c. $-2x + y = 3$
    d. $6y + 2x = 5$

44. Find the equation of the line whose graph contains the points $(-2, -4)$ and $(-6, 5)$.

45. Solve for x: $|3x + 5| = 8$

46. How many different values of x satisfy the equation $|2x + 1| = |4x - 5|$?
    a. 0
    b. 1
    c. 2
    d. 3

For Questions 47 through 50, solve the inequality. Express the solution set using interval notation.

47. $8 - 6x > 50$

48. $4(x + 1) < 5(x + 2)$

49. $3x + 10 \leq -2(x + 15)$

50. $-3(x - 5) - 2 \geq -9(x - 1) + 7x$

51. True or false? The solution set for the inequality $|8x + 3| \geq 3$ is $[0, \infty)$.

52. True or false? The solution set for the inequality $|2x - 3| < 5$ is $(-1, 4)$.

53. Solve the following system using elimination:
$$\begin{cases} x + y = 4 \\ 2x - y = 1 \end{cases}$$

54. Solve the following system using substitution:
$$\begin{cases} 2x + y = 4 \\ 3(y + 9) = 7x \end{cases}$$

55. Solve the following system by graphing:
$$\begin{cases} 3(2x + 3y) = 63 \\ 9(x - 6) = 27y \end{cases}$$

Answers

36. $10x - 2(3 - x) = -3(4 + 2x) - 2x$

$10x - 6 + 2x = -12 - 6x - 2x$

$12x - 6 = -12 - 8x$

$20x = -6$

$x = -\frac{6}{20} = -\frac{3}{10}$

37. $8\left(\frac{1}{4}x - \frac{1}{2}\right) = -2(2 - \frac{1}{4}x)$

$2x - 4 = -4 + \frac{1}{2}x$

$\frac{3}{2}x - 4 = -4$

$\frac{3}{2}x = 0$

$x = 0$

38. $0.8(x + 20) - 4.5 = 0.7(5 + x) - 0.9x$

$0.8x + 16 - 4.5 = 3.5 + 0.7x - 0.9x$

$0.8x + 11.5 = 3.5 - 0.2x$

$x = -8$

39. The slope of the line is 2 and the y-intercept is 3. Hence, the graph of the line $y = 2x + 3$ is given by

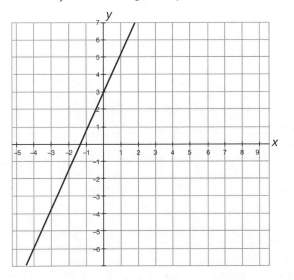

40. The equation $x - 3y = 12$ written in slope-intercept form is $y = \frac{1}{3}x - 4$.
The slope of the line is $\frac{1}{3}$ and the y-intercept is -4. Hence, the graph of
the line $x - 3y = 12$ is given by

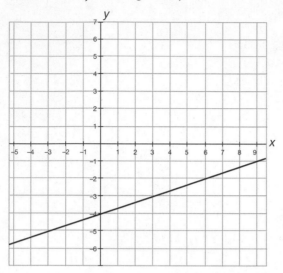

41. The equation $-\frac{5}{3}x - \frac{1}{3}y = -2$ written in slope-intercept form is
$y = -5x + 6$. The slope of the line is -5 and the y-intercept is 6.
Hence, the graph of the line $-\frac{5}{3}x - \frac{1}{3}y = -2$ is given by

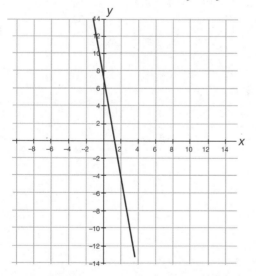

42. a. The y-intercept of the line is $(0,1)$. Since the point $(1,4)$ is also on the
graph, the slope of the line is 3. Thus, substituting $b = 1$ and $m = 3$ into
slope-intercept form $y = mx + b$ of a line yields the equation $y = 3x + 1$.

43. d. The approach is to rewrite each equation in slope-intercept form $y = mx + b$, where m is the slope of the line and b is the y-intercept. Doing so for choice **d** yields the equivalent equation $y = -\frac{1}{3}x + \frac{5}{6}$, which has a negative slope.

44. A point and slope are all that is needed in order to write the equation of the line. The slope m of the line containing the points $(-2, -4)$ and $(-6, 5)$ is

$$m = \frac{5 - (-4)}{-6 - (-2)} = \frac{9}{-4} = -\frac{9}{4}$$

Using the point $(-2, -4)$ as (x_1, y_1) in point-slope form $y - y_1 = m(x - x_1)$ yields the equation of the line, as follows:

$$y - (-4) = -\frac{9}{4}(x - (-2))$$
$$y + 4 = -\frac{9}{4}x - \frac{9}{2}$$
$$y = -\frac{9}{4}x - \frac{17}{2}$$

45. Using the fact that $|a| = b$ if and only if $a = \pm b$, we see that solving the equation $|3x + 5| = 8$ is equivalent to solving $3x + 5 = \pm 8$. We solve these two equations separately, as follows:

$$3x + 5 = -8 \qquad\qquad 3x + 5 = 8$$
$$3x = -13 \qquad\qquad 3x = 3$$
$$x = -\frac{13}{3} \qquad\qquad x = 1$$

Thus, the solutions to the equation are $x = -\frac{13}{3}$ and $x = 1$.

46. c. Note that $|a| = |b|$ if and only if $a = \pm b$. Using this fact, we see that solving the equation $|2x + 1| = |4x - 5|$ is equivalent to solving $2x + 1 = \pm(4x - 5)$. We solve these two equations separately, as follows:

$$2x + 1 = (4x - 5) \qquad 2x + 1 = -(4x - 5)$$
$$6 = 2x \qquad\qquad\quad 2x + 1 = -4x + 5$$
$$\tfrac{1}{3} = x \qquad\qquad\quad 6x = 4$$
$$\qquad\qquad\qquad\qquad x = \tfrac{2}{3}$$

Thus, there are two solutions of the original equation.

47. $8 - 6x > 50$

$$-42 > 6x$$
$$-7 > x$$

So, the solution set for this inequality is $(-\infty, -7)$.

48. $4(x + 1) < 5(x + 2)$

$$4x + 4 < 5x + 10$$
$$-6 < x$$

So, the solution set for this inequality is $(-6, \infty)$.

49. $3x + 10 \leq -2(x + 15)$

$3x + 10 \leq -2x - 30$

$5x \leq -40$

$x \leq -8$

So, the solution set for this inequality is $(-\infty, -8]$.

50. $-3(x - 5) - 2 \geq -9(x - 1) + 7x$

$-3x + 15 - 2 \geq -9x + 9 + 7x$

$-3x + 13 \geq -2x + 9$

$4 \geq x$

So, the solution set for this inequality is $(-\infty, 4]$.

51. False. Note that $|a| \geq c$ if and only if $a \geq c$ or $a \leq -c$. Using this fact, we see that the values of x that satisfy the inequality $|8x + 3| \geq 3$ are precisely those values of x that satisfy either $8x + 3 \geq 3$ or $8x + 3 \leq -3$. We solve these two inequalities separately, as follows:

$8x + 3 \geq 3$ $\qquad\qquad$ $8x + 3 \leq -3$

$8x \geq 0$ $\qquad\qquad\qquad$ $8x \leq -6$

$x \geq 0$ $\qquad\qquad\qquad$ $x \leq -\frac{6}{8} = -\frac{3}{4}$

Thus, the solution set is $[0, \infty) \cup \left(-\infty, -\frac{3}{4}\right]$, not just $[0, \infty)$.

52. True. Note that $|a| < c$ if and only if $-c < a < c$. Using this fact, we see that the values of x that satisfy the inequality $|2x - 3| < 5$ are precisely those values of x that satisfy $-5 < 2x - 3 < 5$. We solve this compound inequality as follows:

$-5 < 2x - 3 < 5$

$-2 < 2x < 8$

$-1 < x < 4$

Thus, the solution set is $(-1, 4)$.

53. First, eliminate y in both equations by simply adding them; then, solve the resulting equation for x, as follows:

$3x = 5$

$x = \frac{5}{3}$

Next, substitute $x = \frac{5}{3}$ into the first equation to determine y, as follows:

$\frac{5}{3} + y = 4$

$y = 4 - \frac{5}{3} = \frac{7}{3}$

Thus, the solution to the system is the ordered pair $\left(\frac{5}{3}, \frac{7}{3}\right)$.

54. First, solve the first equation for y to obtain $y = 4 - 2x$. Then, substitute this expression in for y in the second equation and solve for x, as follows:

$3(y + 9) = 7x$

$3((4 - 2x) + 9) = 7x$

$3(13 - 2x) = 7x$

$39 - 6x = 7x$

$39 = 13x$

$3 = x$

Next, substitute this value of x back into $y = 4 - 2x$ to see that $y = -2$. Thus, the solution to the system is the ordered pair $(3, -2)$.

55. First, solve both equations in the system for y to obtain the equivalent system

$$\begin{cases} y = -\frac{2}{3}x + 7 \\ y = \frac{1}{3}x - 2 \end{cases}$$

Now, graph both lines on the same set of axes, as follows:

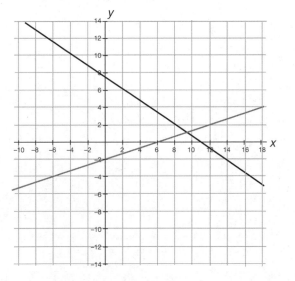

The point of intersection of these two lines, the ordered pair $(9, 1)$, is the solution of the system.

Functions Problems

The types of functions with which we are concerned in single-variable calculus are simply sets of ordered pairs that can be visualized in the Cartesian plane. Such functions are generally described using either algebraic expressions or graphs, and are denoted using letters, such as *f* or *g*. When we want to emphasize the input-output defining relationship of a function, an expression of the form $y = f(x)$ is often used. The arithmetic of real-valued functions is performed using the arithmetic of real numbers and algebraic expressions.

The *domain* of a function can be thought of as the set of all possible *x*-values for which there corresponds an output, *y*. From the graphical viewpoint, an *x*-value belongs to the domain of *f* if an ordered pair with that *x*-value belongs to the graph of *f*. When an algebraic expression is used to describe a function $y = f(x)$, it is convenient to view the domain as the set of all values of *x* that can be substituted into the expression and yield a meaningful output. The *range* of a function is the set of all possible *y*-values attained at some member of the domain.

It is important to gain familiarity with the notion of a function and operations on functions because they constitute the main objects under investigation in the calculus.

Questions

56. Simplify $f(2y-1)$ when $f(x)=x^2+3x-2$.

57. Simplify $f(x+h)-f(x)$ when $f(x)=-(x-1)^2+3$.

58. Compute $(g \circ h)(4)$ when $g(x)=2x^2-x-1$ and $h(x)=x-2\sqrt{x}$.

59. Simplify $(f \circ f \circ f)(2x)$ when $f(x)=-x^2$.

60. Determine the domain of the function $f(x)=\sqrt{-x}$.

61. Determine the domain of the function $g(x)=\dfrac{1}{\sqrt[3]{-1-x}}$.

62. The graph of $f(x)$ is shown here. For how many values of x does $f(x) = 3$?

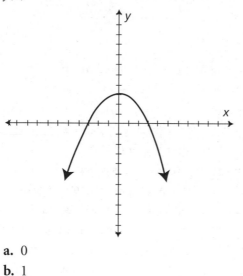

a. 0
b. 1
c. 2
d. 3

63. The graph of $f(x)$ is shown here. For how many values of x does
$f(x) = 0$?

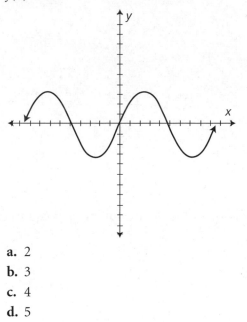

    a. 2
    b. 3
    c. 4
    d. 5

64. The graph of $f(x)$ is shown here. For how many values of x does
$f(x) = 10$?

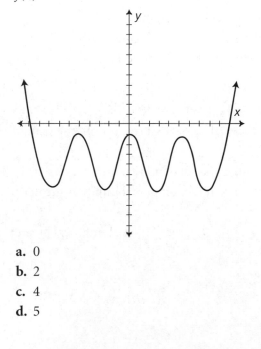

    a. 0
    b. 2
    c. 4
    d. 5

65. If $f(x) = 3x + 2$ and $g(x) = 2x - 3$, what is the value of $g(f(-2))$?
a. -19
b. -11
c. -7
d. -4

66. If $f(x) = 2x + 1$ and $g(x) = x - 2$, what is the value of $f(g(f(3)))$?
a. 1
b. 3
c. 7
d. 11

67. If $f(x) = 6x + 4$ and $g(x) = x^2 - 1$, which of the following is equivalent to $g(f(x))$?
a. $6x^2 - 2$
b. $36x^2 + 16$
c. $36x^2 + 48x + 15$
d. $36x^2 + 48x + 16$

68. What is the range of the function $f(x) = x^2 - 4$?
a. the set of all real numbers excluding 0
b. the set of all real numbers excluding 2 and -2
c. the set of all real numbers greater than or equal to 4
d. the set of all real numbers greater than or equal to -4

69. Which of the following is true of $f(x) = -\frac{1}{2}x^2$?
a. The range of the function is the set of all real numbers less than or equal to 0.
b. The range of the function is the set of all real numbers less than 0.
c. The range of the function is the set of all real numbers greater than or equal to 0.
d. The domain of the function is the set of all real numbers greater than or equal to 0.

70. Which of the following is true of $f(x) = \sqrt{4x - 1}$?

   a. The domain of the function is all real numbers greater than $\frac{1}{4}$, and the range is all real numbers greater than 0.
   b. The domain of the function is all real numbers greater than or equal to $\frac{1}{4}$, and the range is all real numbers greater than 0.
   c. The domain of the function is all real numbers greater than or equal to $\frac{1}{4}$, and the range is all real numbers greater than or equal to 0.
   d. The domain of the function is all real numbers greater than 0, and the range is all real numbers greater than or equal to $\frac{1}{4}$.

For Questions 71 through 77, refer to the functions f and g, both defined on [−5,5], whose graphs are provided here.

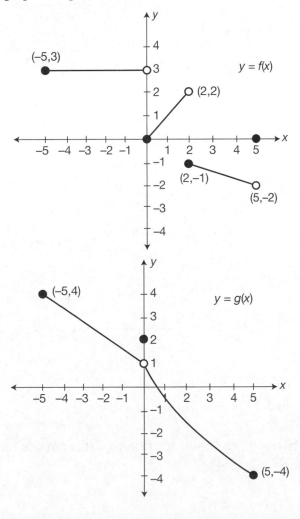

71. True or false? The function f has an inverse.

72. The range of f is which of the following?
   a. $[-2,2]\cup\{3\}$
   b. $(-2,-1]\cup[0,3]$
   c. $(-2,-1]\cup[0,2)\cup\{3\}$
   d. $[-2,2)\cup\{3\}$

73. The range of g is which of the following?
   a. $[-4,4]$
   b. $[-4,2)\cup(2,4]$
   c. $[-4,1)\cup(1,4]$
   d. none of the above

74. $2\cdot f(0)+\left[f(2)\cdot g(5)\right]^{2}=$ _____.
   a. 18
   b. 10
   c. 8
   d. 16

75. $(f\circ g)(0)=$ _____.

   a. $\frac{1}{2}$
   b. -1
   c. 2
   d. undefined

76. $\left(f\left(f\left(f\left(f(5)\right)\right)\right)\right)=$ _____.
   a. 0
   b. -1
   c. 3
   d. undefined

77. True or false? The real number 0 belongs to the domain of the function h defined by $h(x)=\frac{(g\circ g)(x)}{(f\circ f)(x)}$.

78. Which of the following is the domain of the function $f(x) = \dfrac{1}{(2-x)^{\frac{2}{5}}}$?

    a. $(-\infty, 2)$

    b. $(2, \infty)$

    c. $(-\infty, 2) \cup (2, \infty)$

    d. none of the above

79. True or false? There are two x-intercepts of the function
$f(x) = 1 - |2x - 1|$.

80. Determine the inverse function for $f(x) = \dfrac{x-1}{5x+2}$, $x \neq -\frac{2}{5}$.

81. True or false? Assume that the function f has an inverse f^{-1}. If the range of f^{-1} is $[1, \infty)$, then $f(0)$ is not defined.

82. If $f(x) = \sqrt{x^2 - 4x}$, then $f(x+2) = $ _____.

    a. $\sqrt{x^2 - 4x + 2}$

    b. $\sqrt{x^2 - 4x - 4}$

    c. $\sqrt{x^2 - 4}$

    d. $|x - 2|$

83. If $f(x) = \sqrt{-3x}$ and $g(x) = \sqrt{2x^2 + 18}$, then the domain of $g \circ f$ is
_____.

    a. $[0, \infty)$

    b. $(-\infty, 0]$

    c. $\mathbb{R}$

    d. none of the above

84. If $f(x) = \dfrac{1}{2x}$, simplify the expression $\dfrac{f(x+h) - f(x)}{h}$, where $h \neq 0$.

85. Which of the following sequence of shifts would you perform in order to obtain the graph of $f(x) = (x+2)^3 - 3$ from the graph of $g(x) = x^3$?
    a. Shift the graph of $g(x)$ up three units and then left two units.
    b. Shift the graph of $g(x)$ down three units and then right two units.
    c. Shift the graph of $g(x)$ up three units and then right two units.
    d. Shift the graph of $g(x)$ down three units and then left two units.

Answers

56. Substitute the expression $2y - 1$ for every occurrence of x in the definition of the function $f(x)$, and then simplify, as follows:

$$f(2y - 1) = (2y - 1)^2 + 3(2y - 1) - 2$$
$$= 4y^2 - 4y + 1 + 6y - 3 - 2$$
$$= 4y^2 + 2y - 4$$

57. Simplifying $f(x + h)$ requires that you substitute the expression $x + h$ for every occurrence of x in the definition of the function $f(x)$, and then simplify, as follows:

$$f(x + h) = -[(x + h) - 1]^2 + 3 = -[(x + h)^2 - 2(x + h) + 1] + 3$$
$$= -[x^2 + 2hx + h^2 - 2x - 2h + 1] + 3$$
$$= -x^2 - 2hx - h^2 + 2x + 2h - 1 + 3$$
$$= -x^2 - 2hx - h^2 + 2x + 2h + 2$$

Next, in anticipation of simplifying $f(x + h) - f(x)$, we simplify the expression for $f(x) = -(x - 1)^2 + 3$ in order to facilitate combining like terms. Doing so yields

$$f(x) = -(x - 1)^2 + 3 = -(x^2 - 2x + 1) + 3 = -x^2 + 2x - 1 + 3$$
$$= -x^2 + 2x + 2$$

Finally, simplify the original expression $f(x + h) - f(x)$, as follows:

$$f(x + h) - f(x) = (-x^2 - 2hx - h^2 + 2x + 2h + 2) - (-x^2 + 2x + 2)$$
$$= -x^2 - 2hx - h^2 + 2x + 2h + 2 + x^2 - 2x - 2$$
$$= -2hx - h^2 + 2h = h(-h - 2x + 2)$$

58. By definition, $(g \circ h)(4) = g(h(4))$. Observe that $h(4) = 4 - 2\sqrt{4}$ $= 4 - 2(2) = 0$, so $g(h(4)) = g(0) = 2(0)^2 - 0 - 1 = -1$. Thus, we conclude that $(g \circ h)(4) = -1$.

59. By definition, $(f \circ f \circ f)(2x) = f(f(f(2x)))$. Working from the inside outward, we first note that $f(2x) = -(2x)^2 = -4x^2$. Then, $f(f(2x)) =$ $f(-4x^2) = -(-4x^2)^2 = -16x^4$. And finally, $f(f(f(2x))) = f(-16x^4)$ $= -(-16x^4)^2 = -256x^8$. Thus, we conclude that $(f \circ f \circ f)(2x)$ $= -256x^8$.

60. The radicand of an even-indexed radical term (e.g., a square root) must be nonnegative if in the numerator of a fraction and strictly positive if in the denominator of a fraction. For the present function, this restriction takes the form of the inequality $-x \geq 0$, which upon multiplication on both sides by -1 is equivalent to $x \leq 0$. Hence, the domain of the function $f(x) = \sqrt{-x}$ is $(-\infty, 0]$.

61. There is no restriction on the radicand of an odd-indexed radical term (e.g., a cube root) if it is in the numerator of a fraction, whereas the radicand of such a radical term must be nonzero if it occurs in the denominator of a fraction. For the present function, this restriction takes the form of the statement $-1 - x \neq 0$, which is equivalent to $x \neq -1$. Hence, the domain of the function $g(x) = \dfrac{1}{\sqrt[3]{-1-x}}$ is $(-\infty, -1) \cup (-1, \infty)$.

62. b. Draw a horizontal line across the coordinate plane where $f(x) = 3$. This line touches the graph of $f(x)$ in exactly one place. Therefore, there is one value for which $f(x) = 3$.

63. d. $f(x) = 0$ every time the graph touches the x-axis, since the x-axis is the graph of the line $f(x) = 0$. The graph of $f(x)$ touches the x-axis in five places. Therefore, there are five values for which $f(x) = 0$.

64. b. Draw a horizontal line across the coordinate plane where $f(x) = 10$. The arrowheads on the ends of the curve imply that the graph extends upward, without bound, as x tends toward both positive and negative infinity. Hence, this line touches the graph of $f(x)$ in two places. Therefore, there are two values for which $f(x) = 10$.

65. b. Begin with the innermost function: find $f(-2)$ by substituting -2 for x in the function $f(x)$:
$f(-2) = 3(-2) + 2 = -6 + 2 = -4$
Then, substitute that result in for x in $g(x)$.
$g(-4) = 2(-4) - 3 = -8 - 3 = -11$
Thus, $g(f(-2)) = -11$

66. d. Begin with the innermost function: find $f(3)$ by substituting 3 in for x in the function $f(x)$:
$f(3) = 2(3) + 1 = 6 + 1 = 7$
Next, substitute that result in for x in $g(x)$.
$g(7) = 7 - 2 = 5$
Finally, substitute this result in for x in $f(x)$:
$f(5) = 2(5) + 1 = 10 + 1 = 11$
Thus, $f(g(f(3))) = 11$

67. **c.** Begin with the innermost function. You are given the value of $f(x)$: $f(x) = 6x + 4$. Substitute this expression in for x in the equation $g(x)$, and then simplify, as follows:

$g(6x + 4) = (6x + 4)^2 - 1 = 36x^2 + 24x + 24x + 16 - 1 = 36x^2 + 48x + 15$

Therefore, $g(f(x)) = 36x^2 + 48x + 15$.

68. **d.** The domain of a real-valued function is the set of all values that, when substituted in for the variable, produces a meaningful output, while the range of a function is the set of all possible outputs. All real numbers can be substituted in for x in the function $f(x) = x^2 - 4$, so the domain of the function is the set of all real numbers. Since the x term is squared, the smallest value that this term can equal is 0 (when $x = 0$). Therefore, the smallest value that $f(x)$ can attain occurs when $x = 0$. Observe that $f(0) = 0^2 - 4 = -4$. Therefore, the range of $f(x)$ is the set of all real numbers greater than or equal to -4.

69. **a.** Any real number can be substituted for x, so the domain of the function is the set of all real numbers. The range of a function is the set of all possible outputs of the function. Since the x term is squared, then made negative, the largest value that this term can equal is 0 (when $x = 0$). Every other x value will result in a negative value for $f(x)$. Hence, the range of $f(x)$ is the set of all real numbers less than or equal to 0.

70. **c.** The square root of a negative value is imaginary, so the value of $4x - 1$ must be greater than or equal to 0. Symbolically, we have:

$4x - 1 \geq 0$

$4x \geq 1$

$x \geq \frac{1}{4}$

Hence, the domain of $f(x)$ is the set of all real numbers greater than or equal to $\frac{1}{4}$. The smallest value of $f(x)$ occurs at $x = \frac{1}{4}$, and its value is $\sqrt{4\left(\frac{1}{4}\right) - 1} = \sqrt{0} = 0$. So, the range of the function is the set of all real numbers greater than or equal to 0.

71. False. The function f is not one-to-one since the functional value 3 is attained at more than one (in fact, infinitely many) x-values in the domain.

72. c. You must identify all possible *y*-values that are attained within the graph of *f*. The graph of *f* is comprised of three distinct components, each of which contributes an interval of values to the range of *f*. The set of *y*-values corresponding to the bottommost segment is $(-2,1]$; note that -2 is excluded due to the open hole at $(5,-2)$ on the graph, and there is no other *x*-value in $[-5,5]$ whose functional value is -2. Next, the portion of the range corresponding to the middle segment is $[0,2)$; note that 2 is excluded from the range for the same reason -2 is excluded. Finally, the horizontal segment contributes the singleton $\{3\}$ to the range; note that even though there is a hole in the graph at $(0,3)$, there are infinitely many other *x*-values in $[-5,5]$ whose functional value is 3, thereby requiring that it be included in the range. Thus, the range is $(-2,-1]\cup[0,2)\cup\{3\}$.

73. c. The graph of *g* is steadily decreasing from left to right, beginning at the point $(-5,4)$ and ending at $(5,-4)$, with the only gap occurring in the form of a hole at $(0,1)$. Since there is no *x*-value in $[-5,5]$ whose functional value is 1, this value must be excluded from the range. All other values in the interval $[-4,4]$ do belong to the range. Thus, the range is $[-4,1)\cup(1,4]$.

74. d. Using the graphs yields $f(0)=0$, $f(2)=-1$, and $g(5)=-4$. Substituting these values into the given expression yields

$$2\cdot f(0)+\left[f(2)\cdot g(5)\right]^2 = 2(0)+\left[(-1)(-4)\right]^2 = 0+4^2 = 16$$

75. b. Since $g(0) = 2$ and $f(2) = -1$, we have

$$\left(f\circ g\right)(0)= f(g(0))= f(2)=-1$$

76. a. Since $f(5) = 0$ and $f(0) = 0$, we work from the inside outward to obtain

$$f\bigl(f\bigl(f\bigl(f(5)\bigr)\bigr)\bigr)= f\bigl(f\bigl(f(0)\bigr)\bigr)= f\bigl(f(0)\bigr)= f(0)=0$$

77. False. Observe that

$$\left(g\circ g\right)(0)= g(g(0))= g(2)=-2$$
$$\left(f\circ f\right)(0)= f(f(0))= f(0)=0$$

Therefore, computing $h(0)$ would yield the expression $\frac{-2}{0}$, which is not defined. Hence, 0 does not belong to the domain of *h*.

78. c. The radicand of an odd-indexed radical term (e.g., a fifth root) must be nonzero if it occurs in the denominator of a fraction, which is at present the case. Therefore, the restriction takes the form of the statement $2-x\neq 0$, which is equivalent to $x\neq 2$. Thus, the domain is $(-\infty,2)\cup(2,\infty)$.

79. True. The x-intercepts of f are those values of x satisfying the equation $1-|2x-1|=0$, which is equivalent to $|2x-1|=1$. Using the fact that $|a|=b$ if and only if $a=\pm b$, we solve the two equations $2x-1=\pm1$ separately, as follows:

$2x-1=-1$ $\qquad\qquad$ $2x-1=1$

$2x=0$ $\qquad\qquad$ $2x=2$

$x=0$ $\qquad\qquad$ $x=1$

Thus, there are two x-intercepts of the given function.

80. Determining the inverse function for f requires that we solve for x in the expression $y=\frac{x-1}{5x+2}$, as follows:

$y=\frac{x-1}{5x+2}$

$y(5x+2)=x-1$

$5xy+2y=x-1$

$5xy-x=-2y-1$

$x(5y-1)=-2y-1$

$x=\frac{-2y-1}{5y-1}$

Now, we conclude that the function $f^{-1}(y)=\frac{-2y-1}{5y-1}$, $y\neq\frac{1}{5}$ is the inverse function of f.

81. True. Remember that the domain of f is equal to the range of f^{-1}.

Therefore, since 0 does not belong to the range of f^{-1}, it does not belong to the domain of f, so $f(0)$ is undefined.

82. c. Simplify the given expression, as follows:

$$f(x+2)=\sqrt{(x+2)^2-4(x+2)}=\sqrt{x^2+4x+4-4x-8}=\sqrt{x^2-4}$$

83. b. The domain of $g\circ f$ consists of only those values of x for which the quantity $f(x)$ is defined (that is, x belongs to the domain of f) *and* for which $f(x)$ belongs to the domain of g. For the present scenario, the domain of f consists of only those x-values for which $-3x\geq0$, which is equivalent to $x\leq0$. Since the domain of $g(x)=\sqrt{2x^2+18}$ is the set of all real numbers, it follows that all x values in the interval $(-\infty,0]$ are permissible inputs in the composition function $(g\circ f)(x)$, and that, in fact, these are the only permissible inputs. Therefore, the domain of $g\circ f$ is $(-\infty,0]$.

84. Simplify the given expression, as follows:

$$\frac{f(x+h)-f(x)}{h}=\frac{\frac{1}{2(x+h)}-\frac{1}{2x}}{h}=\frac{\frac{x-(x+h)}{2x(x+h)}}{h}=\frac{x-x-h}{2hx(x+h)}=\frac{-h}{2hx(x+h)}=\frac{-1}{2x(x+h)}$$

85. d. The graph of $y = f(x + h)$ is obtained by shifting the graph of $y = f(x)$ to the left h units if $h > 0$, and to the right h units if $h < 0$. Also, the graph of $y = g(x) + k$ is obtained by shifting the graph of $y = g(x)$ up k units if $k > 0$, and down k units if $k < 0$. Using both of these facts, in conjunction, shows that the graph of $f(x) = (x + 2)^3 - 3$ is obtained by shifting the graph of $g(x) = x^3$ down three units and left two units (in either order).

4

Polynomial Problems

A _polynomial function_ is of the form $p(x) = a_n x^n + a_{n-1} x^{n-1} + \ldots + a_1 x + a_0$, where $a_0, \ldots, a_n$ are real numbers with $a_n \neq 0$, and n is a nonnegative integer; we say that $p(x)$ has nth degree. The zeros of a polynomial are its x-intercepts and are typically found using the factored version of $p(x)$. An nth degree polynomial can have at most n distinct x-intercepts and at most $n - 1$ turning points.

Questions

For Questions 86 through 93, refer to the graph of the following fourth degree polynomial function $y = p(x)$.

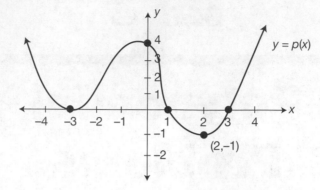

86. The zeros of $p(x)$ are $x = $ _____.
   a. $-3, 0, 2$
   b. $-3, 1, 3$
   c. $-3, 0, 1, 2, 3$
   d. none of the above

87. Which of the following three statements are true?

   I. $(x + 3)^2$ divides evenly into the expression defining the function $p(x)$.

   II. $(x - 3)^2$ divides evenly into the expression defining the function $p(x)$.

   III. $(x + 1)$ divides evenly into the expression defining the function $p(x)$.
   a. I only
   b. I and II only
   c. I, II, and III
   d. none of the above

88. Which of the following is the range of $p(x)$?
   a. $\mathbb{R}$
   b. $[-1, 4]$
   c. $[-1, \infty)$
   d. $[-4, 4]$

89. Which of the following is the domain of $p(x)$?

    a. $\mathbb{R}$

    b. $[-1,4]$

    c. $[-1,\infty)$

    d. $[-4,4]$

90. Which of the following is the solution set for the inequality $-1 < p(x) \leq 0$?

    a. $(1,3)$

    b. $[1,3] \cup \{-3\}$

    c. $(1,2) \cup (2,3)$

    d. $[1,2) \cup (2,3] \cup \{-3\}$

91. Determine the solution set for the inequality $p(x) > 0$.

92. True or false? There exists an x-value for which $p(x) = 1,500$.

93. On what intervals is the graph of $y = p(x)$ increasing?

    a. $(-3,0) \cup (2,\infty)$

    b. $(-3,0) \cup (3,\infty)$

    c. $(-\infty,-3) \cup (0,1)$

    d. $(-\infty,-3) \cup (0,2)$

94. Which of the following is the solution set for the inequality $2(x-3)(x+1)^2(x+2) < 0$?

    a. $(-2,3)$

    b. $[-2,3]$

    c. $(-2,-1) \cup (-1,3)$

    d. none of the above

95. Determine the solution set for the inequality $x^4 + 16 \leq 8x^2$.

96. True or false? There does not exist a third-degree polynomial that is increasing on the entire set of real numbers.

97. True or false? There exists a fifth-degree polynomial that does not pass through the origin and that is decreasing on the entire set of real numbers.

98. The polynomial $(x-6)(x-3)(x-1)$ is equal to which of the following?
   a. $x^3 - 9x - 18$
   b. $x^3 - 8x^2 + 27x - 18$
   c. $x^3 - 10x^2 - 9x - 18$
   d. $x^3 - 10x^2 + 27x - 18$

99. The polynomial $2x^3 + 8x^2 - 192x$ is equal to which of the following?
   a. $2(x-8)(x+12)$
   b. $2x(x-8)(x+12)$
   c. $x(2x-8)(x+24)$
   d. $2x(x+16)(x-12)$

100. Which of the following is a solution of the equation $x(x-1)(x+1) = 27 - x$?
   a. -9
   b. -1
   c. 3
   d. 1

101. The cube of a number minus twice its square is equal to 80 times the number. If the number is greater than 0, what is the number?
   a. 4
   b. 5
   c. 8
   d. 10

102. The product of three consecutive positive integers is equal to 56 more than the cube of the first integer. What is largest of these integers?
   a. 3
   b. 4
   c. 5
   d. 6

103. True or false? There exists a polynomial whose graph never intersects the x-axis.

104. True or false? It is possible to construct a sixth-degree polynomial that has only three x-intercepts and whose graph never extends above the x-axis.

105. True or false? A polynomial must have a y-intercept.

Answers

86. b. The zeros of a polynomial are its x-intercepts, which are -3, 1, and 3.

87. a. Since -3, 1, and 3 are all zeros of $p(x)$, we know that at the very least, each of $(x+3), (x-1),$ and $(x-3)$ must divide into the expression defining $p(x)$. Moreover, no other linear factor can divide into $p(x)$, because this would imply the existence of a fourth zero, which does not exist. Now, since it is assumed that $p(x)$ is of fourth degree, we know that exactly one of these factors must occur twice. The only factor that fulfills this condition is $(x+3)$ because the graph of $p(x)$ is tangent to the x-axis at $x = -3$. Hence, we conclude that $(x+3)^2$ divides evenly into the expression defining the function $p(x)$.

88. c. The lowest point on the graph of $y = p(x)$ occurs at $(2,-1)$, so the smallest possible y-value attained is -1. Further, every real number greater than -1 is also attained at some x-value. Hence, the range is $[-1, \infty)$.

89. a. The domain of any polynomial function is the set of real numbers, $\mathbb{R}$, because any real number can be substituted for x in $p(x)$ and yield another real number.

90. d. We must identify the x-values of the portion of the graph of $y = p(x)$ that lies between the horizontal lines $y = -1$ and $y = 0$ (i.e., the x-axis). Once this is done, we exclude the x-values of the points where the graph of $y = p(x)$ intersects the line $y = -1$ (because of the strict inequality), and we include those x-values of the points where the graph of $y = p(x)$ intersects the line $y = 0$. Doing so yields the set $[1, 2) \cup (2, 3] \cup \{-3\}$.

91. We must identify the x-values of the portion of the graph of $y = p(x)$ that lies strictly above the x-axis. Using the graph, we see that this corresponds to the set $(-\infty, -3) \cup (-3, 1) \cup (3, \infty)$.

92. True. The graph of $y = p(x)$ extends upward without bound, as evidenced by the arrowheads affixed to the extremities of the graph. The graph will therefore eventually intersect the line $y = 1{,}500$ twice, so there exists an x-value for which $p(x) = 1{,}500$.

93. a. We must identify the intervals in the domain of $p(x)$ on which the graph of $y = p(x)$ rises from left to right. This happens precisely on the intervals $(-3, 0) \cup (2, \infty)$.

94. c. The strategy is to determine the x-values that make the expression on the left side equal to zero, and then to assess the sign of the expression on the left side on each subinterval formed using these zeros. To this end, observe that the zeros are $x = -2, -1$, and 3. Now, we form a number line, choose a real number in each of the subintervals, and record the sign of the expression above each, as follows:

As such, the solution set is $(-2, -1) \cup (-1, 3)$ because those are the intervals where the inequality is less than zero.

95. First, bring all terms to the left side and then factor, as follows:

$$x^4 + 16 \le 8x^2$$

$$x^4 - 8x^2 + 16 \le 0$$

$$\left(x^2 - 4\right)^2 \le 0$$

Observe that the expression on the left side of the inequality, being the square of a binomial, is always nonnegative. Hence, the only x-values that satisfy the inequality are those for which $x^2 - 4 = 0$. This equation is equivalent to $(x - 2)(x + 2) = 0$, which has solutions $x = \pm 2$. These are, in fact, the only solutions of the original inequality.

96. False. For instance, the graph of the polynomial $p(x) = x^3$ is increasing on the entire set of real numbers.

97. True. For instance, the graph of the polynomial $p(x) = -(x-1)^5$ does not pass through the origin and is decreasing on the entire set of real numbers.

98. d. Begin by multiplying the first two terms: $(x - 6)(x - 3) = x^2 - 3x - 6x + 18 = x^2 - 9x + 18$. Multiply $(x^2 - 9x + 18)$ by $(x - 1)$: $(x^2 - 9x + 18)(x - 1) = x^3 - 9x^2 + 18x - x^2 + 9x - 18 = x^3 - 10x^2 + 27x - 18$.

99. b. The largest constant common to each term is 2, and x is the largest common variable. Factor out $2x$ from every term: $2x^3 + 8x^2 - 192x = 2x(x^2 + 4x - 96)$. Next, factor $x^2 + 4x - 96$ into $(x - 8)(x + 12)$. Thus, the factored version of $2x^3 + 8x^2 - 192x$ is $2x(x - 8)(x + 12)$.

100. c. First, multiply the terms on the left side of the equation: $x(x - 1) = x^2 - x$ and subsequently, $(x^2 - x)(x + 1) = x^3 + x^2 - x^2 - x = x^3 - x$. Therefore, $x^3 - x = 27 - x$. Next, add x to both sides of the equation to obtain $x^3 - x + x = 27 - x + x$, which is equivalent to $x^3 = 27$. Finally, the cube root of 27 is 3, so the solution of $x(x - 1)(x + 1) = 27 - x$ is $x = 3$.

101. d. If the number is x, then the cube of the number is x^3. Twice the square of the number is $2x^2$. The difference in those values is equal to 80 times the number ($80x$). Therefore, $x^3 - 2x^2 = 80x$. Subtract $80x$ from both sides of the equation, and factor the polynomial, as follows:

$x^3 - 2x^2 = 80x$

$x^3 - 2x^2 - 80x = 0$

$x(x^2 - 2x - 80) = 0$

$x(x + 8)(x - 10) = 0$

Set each factor equal to 0 to find that the values of x that make the equation true are 0, –8, and 10. Since 10 is the only number among these values that is greater than 0, it is the solution we seek.

102. d. If the first integer is x, then the second integer is $(x + 1)$ and the third integer is $(x + 2)$. The product of these integers is

$$x(x + 1)(x + 2) = (x^2 + x)(x + 2)$$
$$= x^3 + 2x^2 + x^2 + 2x$$
$$= x^3 + 3x^2 + 2x$$

This polynomial is equal to the cube of the first integer, x^3, plus 56. Therefore,

$x^3 + 3x^2 + 2x = x^3 + 56$

Move all terms to one side of the equation, combining like terms to obtain the equivalent equation $3x^2 + 2x - 56 = 0$. Now, factor the polynomial: $3x^2 + 2x - 56 = (3x + 14)(x - 4)$. Set each term equal to 0 to see that the solutions of the equation are $x = -\frac{14}{3}$ and $x = 4$. Since we are looking for a positive integer, we use 4. Hence, using 4 as the first of the consecutive integers means that 5 is the second integer and 6 is the third, and largest, integer.

103. True. This is true for many polynomials. For instance, the graph of $p(x) = x^2 + 1$ lies strictly above the x-axis.

104. True. For instance, consider $p(x) = -(x - 1)^2(x - 2)^2(x - 3)^2$. The x-intercepts are 1, 2, and 3, and since the expression involves three squared binomials, the product of which is multiplied by –1, the expression for $p(x)$ is never positive; hence, its graph can never extend above the x-axis.

105. True. The domain of a polynomial function is the set of all real numbers. So, in particular, 0 belongs to the domain and the graph must intersect the line $x = 0$, which is the y-axis.

Rational Expression Problems

A *rational expression* is an algebraic expression of the form $\frac{p(x)}{q(x)}$, where p and q are polynomials. The arithmetic of rational expressions resembles the arithmetic of fractions, where the factors of $p(x)$ and $q(x)$ play the role of the prime factors.

A *rational function* is of the form $f(x) = \frac{p(x)}{q(x)}$, where p and q are polynomial functions. The domain of a rational function is the set of all real numbers for which the denominator is not equal to zero, and the function has vertical asymptotes at every x-value excluded from the domain that does not make the numerator equal to zero. If an x-value, say $x = a$, makes both the numerator and the denominator equal to zero, then there is an open hole in the graph at $x = a$.

The function has a horizontal asymptote if and only if the degree of the numerator (i.e., the highest power of x) is less than or equal to the degree of the denominator. If the degree of the numerator is less than that of the denominator, then the horizontal asymptote is $y = 0$ (i.e., the x-axis), while if the degrees are the same, then the horizontal asymptote is $y = c$, where c is the quotient of the coefficients of terms of highest degree in the numerator and denominator. The function has an oblique asymptote if and only if the degree of the numerator is exactly one more than the degree of the denominator; in such case, the asymptote is of the form $y = ax + b$ and is obtained by dividing the numerator by the denominator using the method of long division. If the degree of the numerator exceeds that of the denominator by more than one, then the function has neither a horizontal nor an oblique asymptote.

Questions

106. The fraction $\dfrac{x^2 + 8x}{x^3 - 64x}$ is equivalent to which of the following?

   a. $\dfrac{1}{x-8}$

   b. $\dfrac{x}{x-8}$

   c. $\dfrac{x+8}{x-8}$

   d. $x-8$

107. The fraction $\dfrac{2x^2 + 4x}{4x^3 - 16x^2 - 48x}$ is equivalent to which of the following?

   a. $\dfrac{x+2}{x-6}$

   b. $\dfrac{x}{(x+2)(x+6)}$

   c. $\dfrac{1}{2x-12}$

   d. $\dfrac{x+2}{4x(x-6)}$

108. Which of the following makes the fraction $\dfrac{x^2 + 11x + 30}{4x^3 + 44x^2 + 120x}$ undefined?

   a. -6

   b. -4

   c. -3

   d. -2

109. Determine the values of n, if any, that satisfy the equation
$\dfrac{1}{n} + \dfrac{1}{n+1} = \dfrac{-1}{n(n+1)}$.

110. Determine the values of p, if any, that satisfy the equation
$\dfrac{1}{p+4} + \dfrac{1}{p-4} = \dfrac{p^2 - 48}{p^2 - 16}$.

111. Determine the values of t, if any, that satisfy the equation $\dfrac{1}{\frac{2}{t-3} + \frac{1}{t+3}} = 0$.

112. If $f(x) = \dfrac{1}{x-3}$, simplify the expression $\dfrac{f(x+h) - f(x)}{h}$.

113. Determine the solution set for the inequality $\dfrac{(x-1)(x+2)}{(x+3)^2} \leq 0$.

114. Determine the solution set for the inequality $\dfrac{x^2 + 9}{x^2 - 2x - 3} > 0$.

115. Determine the solution set for the inequality $\dfrac{-x^2 - 1}{6x^4 - x^3 - 2x^2} \geq 0$.

116. Determine the solution set for the inequality $\dfrac{\frac{2}{x^2} - \frac{1}{x-1}}{\frac{1}{x+3} - \frac{4}{x^2}} \geq 0$.

117. True or false? A rational function must have either an x-intercept or a y-intercept.

118. Which of the following are characteristics of the graph of
$f(x) = 2 - \frac{x^2 + 1}{x - 1}$?

   I. The function is equivalent to the linear function $g(x) = 2 - (x+1)$ with a hole at $x = 1$.

   II. There is one vertical asymptote, no horizontal asymptote, and an oblique asymptote.

   III. There is one x-intercept and one y-intercept.

   a. I only

   b. II only

   c. II and III only

   d. I and III only

119. Which of the following are characteristics of the graph of
$f(x) = \frac{(2 - x)^2 (x + 3)}{x(x - 2)^2}$?

   I. The graph has a hole at $x = 2$.

   II. $y = 1$ is a horizontal asymptote and $x = 0$ is a vertical asymptote.

   III. There is one x-intercept and one y-intercept.

   a. I and III only

   b. I and II only

   c. I only

   d. none of these choices

120. The domain of $f(x) = \frac{2x}{x^3 - 4x}$ is

   a. $(-\infty, -2) \cup (2, \infty)$.

   b. $(-\infty, 2) \cup (2, \infty)$.

   c. $(-\infty, -2) \cup (-2, 0) \cup (0, 2) \cup (2, \infty)$.

   d. $(-\infty, -2) \cup (-2, 2) \cup (2, \infty)$.

121. Which of the following are the vertical and horizontal asymptotes, if any, for the function $f(x) = \frac{(x-3)(x^2-16)}{(x^2+9)(x-4)}$?

    a. $x = -3$, $x = 4$

    b. $x = -3$, $x = 4$, $y = 1$

    c. $x = 4$, $y = 1$

    d. $y = 1$

122. The range of the function $f(x) = \frac{2x+1}{1-x}$ is which of the following?

    a. $(-\infty, 1) \cup (1, \infty)$

    b. $(-\infty, -2) \cup (-2, \infty)$

    c. $\left(-\infty, -\frac{1}{2}\right) \cup \left(-\frac{1}{2}, \infty\right)$

    d. $\mathbb{R}$

123. True or false? There exists a rational function whose range is $(1, \infty)$.

124. True or false? There exists a rational function with horizontal asymptote $y = -2$ and vertical asymptotes $x = -1$, $x = 0$, and $x = 1$.

125. True or false? A rational function can have both a horizontal asymptote and an oblique asymptote.

Answers

106. a. Factor the numerator and denominator, as follows:

$x^2 + 8x = x(x + 8)$

$x^3 - 64x = x(x^2 - 64) = x(x - 8)(x + 8)$

Cancel the x terms and the $(x + 8)$ terms in the numerator and denominator, leaving 1 in the numerator and $(x - 8)$ in the denominator.

107. c. Factor the numerator and denominator, as follows:

$2x^2 + 4x = 2x(x + 2)$

$4x^3 - 16x^2 - 48x = 4x(x^2 - 4x - 12) = 4x(x - 6)(x + 2)$

Cancel the $2x$ term in the numerator with the $4x$ term in denominator, leaving 2 in the denominator. Cancel the $(x + 2)$ terms in the numerator and denominator, leaving $2(x - 6) = 2x - 12$ in the denominator and 1 in the numerator.

108. a. A rational expression is undefined for any x-value that makes its denominator equal to zero. Factor the polynomial in the denominator and set each factor equal to zero to find the values that make the fraction undefined, as follows:

$4x^3 + 44x^2 + 120x = 4x(x^2 + 11x + 30) = 4x(x + 5)(x + 6)$

$4x = 0, x = 0; x + 5 = 0, x = -5; x + 6 = 0, x = -6$

The fraction is undefined when x is equal to $-6, -5$, or 0.

109. Multiply both sides of the equation by the lowest common denominator (LCD) $n(n + 1)$. Then, solve the resulting equation, as follows:

$\frac{1}{n} + \frac{1}{n+1} = \frac{-1}{n(n+1)}$

$n(n+1)\left[\frac{1}{n} + \frac{1}{n+1}\right] = n(n+1)\left[\frac{-1}{n(n+1)}\right]$

$\frac{n(n+1)}{n} + \frac{n(n+1)}{n+1} = \frac{-n(n+1)}{n(n+1)}$

$(n+1) + n = -1$

$2n = -2$

$n = -1$

But note that substituting this value for n in the original equation renders two of the terms undefined. Hence, this value cannot satisfy the equation, so we conclude that the equation has no solution.

110. Rewrite all terms on the left side of the equation using the LCD $(p+4)(p-4)$, combine the fractions into a single fraction, and simplify, as follows:

$$\frac{1}{p+4} + \frac{1}{p-4} - \frac{p^2-48}{p^2-16} = 0$$

$$\frac{(p-4)+(p+4)-\left(p^2-48\right)}{(p+4)(p-4)} = 0$$

$$\frac{-\left(p^2-2p-48\right)}{(p+4)(p-4)} = 0$$

$$\frac{(p-8)(p+6)}{(p+4)(p-4)} = 0$$

Since there are no factors common to the numerator and denominator, the values of p that make the numerator equal to zero are the solutions of the original equation. These values are $p = -6$ and 8.

111. Rewrite both fractions in the denominator of the fraction on the left side of the equation using the LCD $(t-3)(t+3)$, combine the fractions into a single fraction, and simplify, as follows:

$$\frac{1}{\frac{2}{t-3}+\frac{1}{t+3}} = 0$$

$$\frac{1}{\frac{2(t+3)+(t-3)}{(t-3)(t+3)}} = 0$$

$$\frac{1}{\frac{3t+3}{(t-3)(t+3)}} = 0$$

$$\frac{(t-3)(t+3)}{3(t+1)} = 0$$

Now, even though the values $t = -3$ and $t = 3$ both satisfy the last equation in the preceding string, they do not satisfy the original equation because they each make the denominator of a fraction within the expression equal to zero. Therefore, there is no solution of this equation.

(Note: In retrospect, note that the original equation is of the form $\frac{1}{b} = 0$. Since a fraction equals zero if and only if its numerator equals zero, we could have immediately concluded that the equation had no solution.)

112. $\dfrac{f(x+h)-f(x)}{h} = \dfrac{\frac{1}{(x+h)-3} - \frac{1}{x-3}}{h} = \dfrac{(x-3)-(x+h-3)}{h(x+h-3)(x-3)} = \dfrac{x-3-x-h+3}{h(x+h-3)(x-3)}$

$\qquad\qquad\qquad = \dfrac{-h}{h(x+h-3)(x-3)} = \dfrac{-1}{(x+h-3)(x-3)}$

113. The strategy is to determine the x-values that make the expression on the left side of the equation equal to zero or undefined. Then, we assess the sign of the expression on the left side on each subinterval formed using these values. To this end, observe that these values are $x = -3, -2,$ and 1. Now, we form a number line, choose a real number in each of the duly formed subintervals, and record the sign of the expression above each, as follows:

```
    +      +      −      +
 ←——+——————+——————+——————→
   −3     −2      1
```

Since the $\leq$ inequality includes "equals," we include those values from the number line that make the numerator equal to zero. Hence, the solution set is $[-2,1]$.

114. First, we must make certain that the numerator and the denominator are both completely factored and that all common terms are canceled, as follows:

$$\dfrac{x^2+9}{x^2-2x-3} = \dfrac{x^2+9}{(x-3)(x+1)}$$

Now, the strategy is to determine the x-values that make this expression equal to zero or undefined. Then, we assess the sign of the expression on the left side on each subinterval formed using these values. To this end, observe that these values are $x = -1$ and 3. Now, we form a number line, choose a real number in each of the duly formed subintervals, and record the sign of the expression above each, as follows:

```
    +       −       +
 ←——+———————+———————→
   −1       3
```

Since the $>$ inequality does not include "equals," we do not include those values from the number line that make the numerator equal to zero. Hence, the solution set is $(-\infty,-1)\cup(3,\infty)$.

115. First, we must make certain that the numerator and the denominator are both completely factored and that all common terms are canceled, as follows:

$$\frac{-x^2 - 1}{6x^4 - x^3 - 2x^2} = \frac{-(x^2 + 1)}{x^2(6x^2 - x - 2)} = \frac{-(x^2 + 1)}{x^2(2x + 1)(3x - 2)}$$

Now, the strategy is to determine the x-values that make this expression equal to zero or undefined. Then, we assess the sign of the expression on the left side on each subinterval formed using these values. To this end, observe that these values are $x = -\frac{1}{2}$, 0, and $\frac{2}{3}$.

Now, we form a number line, choose a real number in each of the duly formed subintervals, and record the sign of the expression above each, as follows:

Since the $\geq$ inequality includes "equals," we would include those values from the number line that make the numerator equal to zero. Since none of these values make the numerator equal to zero, we conclude that the solution set is $\left(-\frac{1}{2}, 0\right) \cup \left(0, \frac{2}{3}\right)$.

116. First, we must simplify the complex fraction on the left side of the inequality, as follows:

$$\frac{\dfrac{2}{x^2}-\dfrac{1}{x-1}}{\dfrac{1}{x+3}-\dfrac{4}{x^2}}=\frac{\dfrac{2(x-1)-x^2}{x^2(x-1)}}{\dfrac{x^2-4(x+3)}{x^2(x+3)}}=\frac{-\left(x^2-2x+1\right)}{x^2(x-1)}\cdot\frac{x^2(x+3)}{x^2-4x-12}$$

$$=\frac{-(x-1)^2}{x^2(x-1)}\cdot\frac{x^2(x+3)}{(x-6)(x+2)}=-\frac{(x-1)(x+3)}{(x-6)(x+2)}$$

So, the original inequality can be written as $-\dfrac{(x-1)(x+3)}{(x-6)(x+2)}\geq 0$, or

equivalently (upon multiplication by –1 on both sides), $\dfrac{(x-1)(x+3)}{(x-6)(x+2)}\leq 0$.

Now, the strategy is to determine the x-values that make this expression equal to zero or undefined, including the values that make any factors common to both numerator and denominator equal to zero. Then, we assess the sign of the expression on the left side on each subinterval formed using these values. To this end, observe that these values are $x=$ $-3, -2, 0, 1,$ and 6. Now, we form a number line, choose a real number in each of the duly formed subintervals, and record the sign of the expression above each, as follows:

Since the $\geq$ inequality includes "equals," we would include those values from the number line that make the numerator equal to zero.

Therefore, the solution set is $[-3,-2)\cup[1,6)$.

117. False. For instance, the function $f(x)=\frac{1}{x}$ crosses neither the x-axis nor the y-axis.

118. b. First, note that x^2+1 does not factor, so $f(x)$ cannot be simplified any further. This also means that $x=1$ is a vertical asymptote for the graph of f. Since the degree of the numerator of the fraction is exactly one more than that of the denominator, we can conclude that the graph has no horizontal asymptote, but does have an oblique asymptote. Hence, II is a characteristic of the graph of f.

Next, while there is a y-intercept, $(0,3)$, there is no x-intercept. To see this, we must consider the equation $f(x)=0$, which is equivalent to

$$2-\frac{x^2+1}{x-1}=\frac{2(x-1)-\left(x^2+1\right)}{x-1}=\frac{-\left(x^2-2x+3\right)}{x-1}=0$$

The x-values that satisfy such an equation are those that make the numerator equal to zero *and* do not make the denominator equal to zero. Since the numerator does not factor, we know that $(x-1)$ is not a factor of it, so we need only solve the equation $x^2-2x+3=0$. Using the quadratic formula yields

$$x=\frac{-(-2)\pm\sqrt{(-2)^2-4(1)(3)}}{2(1)}=\frac{2\pm\sqrt{-8}}{2}=1\pm i\sqrt{2}$$

Since the solutions are imaginary, we conclude that there are no x-intercepts. Hence, III is not a characteristic of the graph of f.

119. b. The expression for f can be simplified as follows:

$$\frac{(2-x)^2(x+3)}{x(x-2)^2}=\frac{(-(x-2))^2(x+3)}{x(x-2)^2}=\frac{(x-2)^2(x+3)}{x(x-2)^2}=\frac{x+3}{x}$$

Since $x=2$ makes both the numerator and the denominator of the unsimplified expression equal to zero, there is a hole in the graph of f at this value. So, I holds.

Next, since the degrees of the numerator and the denominator are equal, there is a horizontal asymptote given by $y=1$ (since the quotient of the coefficients of the terms of highest degree in the numerator and the denominator is $1\div1=1$). Moreover, $x=0$ makes the denominator equal to zero, but does not make the numerator equal to zero; hence, it is a vertical asymptote, and II holds.

Finally, since $x=0$ is a vertical asymptote, the graph of f cannot intersect it. Hence, there is no y-intercept, so III does not hold.

120. c. The domain of a rational function is the set of all real numbers that do not make the denominator equal to zero. For this function, the values of x that must be excluded from the domain are the solutions of the equation $x^3 - 4x = 0$. Factoring the left side yields the equivalent equation

$$x^3 - 4x = x\left(x^2 - 4\right) = x(x-2)(x+2) = 0$$

the solutions of which are $x = -2, 0,$ and 2. Hence, the domain is $(-\infty, -2) \cup (-2, 0) \cup (0, 2) \cup (2, \infty)$.

121. d. First, simplify the expression for $f(x)$ as follows:

$$\frac{(x-3)\left(x^2 - 16\right)}{\left(x^2 + 9\right)(x-4)} = \frac{(x-3)(x-4)(x+4)}{\left(x^2 + 9\right)(x-4)} = \frac{(x-3)(x+4)}{x^2 + 9} = \frac{x^2 + x - 12}{x^2 + 9}$$

While there is a hole in the graph of f at $x = 4$, there is no x-value that makes the denominator of the simplified expression equal to zero. Hence, there is no vertical asymptote. But, since the degrees of the numerator and the denominator are equal, there is a horizontal asymptote given by $y = 1$ (since the quotient of the coefficients of the terms of highest degree in the numerator and the denominator is $2 \div 2 = 1$).

122. b. The graph of f has a vertical asymptote at $x = 1$ and a horizontal asymptote at $y = -2$. Since the graph follows the vertical asymptote up to positive infinity as x approaches $x = 1$ from the left and down to negative infinity as x approaches $x = 1$ from the right, and it does not cross the horizontal asymptote, we conclude that the graph attains all y-values except -2. Hence, the range is $(-\infty, -2) \cup (-2, \infty)$.

123. True. For instance, the function $f(x) = \frac{1}{x^2} + 1$ is such an example.

124. True. The denominator of such a function must include the factors $(x+1)$, x, and $(x-1)$, and these factors cannot belong to the numerator. Moreover, in order to have a horizontal asymptote at $y = -2$, the degrees of the numerator and denominator must be equal and the quotient of the coefficients of the terms of highest degree must be -2. An example of such a function satisfying all of these conditions is $f(x) = \frac{-2x^3 + 1}{x(x+1)(x-1)}$.

125. False. A rational function has a horizontal asymptote if and only if the degree of the numerator is less than or equal to the degree of the denominator, whereas a rational function has an oblique asymptote if and only if the degree of the numerator is exactly one more than the degree of the denominator. These two conditions cannot hold simultaneously.

6

Exponential and Logarithmic Functions Problems

An *exponential function* is of the form $f(x) = a^x$, where the base a is a positive real number not equal to 1. The domain of such a function is the set of all real numbers, and the range is $(0, \infty)$. The usual exponent rules apply when performing various arithmetic operations on such functions.

The function $f(x) = a^x$ is one-to-one for all such choices of a, and hence has an inverse called a *logarithmic function*, denoted by $f^{-1}(x) = \log_a x$. The domain of this function is $(0, \infty)$, and its range is the set of all real numbers. When the base a is equal to the irrational number e, we write $\ln x$ instead of $\log_e x$ to denote the fact that this is the *natural logarithm*.

Basic Properties of Logarithms

Some basic properties of logarithms are as follows:

1. $\log_a x$ is the exponent y such that $a^y = x$; that is, $\log_a x = y$ if and only if $a^y = x$.
2. $\log_a(xy) = \log_a x + \log_a y$
3. $\log_a\left(\frac{x}{y}\right) = \log_a x - \log_a$
4. $\log_a(x^y) = y\log_a x$
5. If $\log_a x = \log_a y$, then $x = y$.
6. $a^{\log_a x} = x = \log_a(a^x)$
7. If $x < y$, then $\ln x < \ln y$.

Questions

126. $\log_3 27 = $ _____

127. $\log_3\left(\frac{1}{9}\right) = $ _____

128. $\log_{\frac{1}{2}} 8 = $ _____

129. $\log_7 \sqrt{7} = $ _____

130. $\log_5 1 = $ _____

131. $\log_{16} 64 = $ _____

132. If $\log_6 x = 2$, then $x = $ _____ .

133. If $5\sqrt{a} = x$, then $\log_a x = $ _____ .

134. $\log_3\left(3^4 \cdot 9^3\right) = $ _____

135. If $5^{3x-1} = 7$, then $x = $ _____ .

136. If $\log_a x = 2$ and $\log_a y = -3$, then $\log_a\left(\frac{x}{y^3}\right) = $ _____ .

137. $3^{\log_3 2} = $ _____

138. $\log_a(a^x) = $ _____

139. If $3\ln\left(\frac{1}{x}\right) = \ln 8$, then $x = $ _____.

140. $e^{-\frac{1}{2}\ln 3} = $ _____

141. If $\ln x = 3$ and $\ln y = 2$, then $\ln\left(\frac{e^2 y}{\sqrt{x}}\right) = $ _____.

142. If $e^x = 2$ and $e^y = 3$, then $e^{3x-2y} = $ _____.

143. The range of the function $f(x) = 1 - 2e^x$ is which of the following?
    a. $(-\infty, 1]$
    b. $(-\infty, 1)$
    c. $(1, \infty)$
    d. $[1, \infty)$

144. The range of the function $g(x) = \ln(2x - 1)$ is which of the following?
    a. $\left(-\infty, -\frac{1}{2}\right)$
    b. $\left[\frac{1}{2}, \infty\right)$
    c. $\left(\frac{1}{2}, \infty\right)$
    d. $\mathbb{R}$

145. Which of the following, if any, are x-intercepts of the function
    $f(x) = \ln\left(x^2 - 4x + 4\right)$?
    a. $(1, 0)$
    b. $(3, 0)$
    c. both **a** and **b**
    d. neither **a** nor **b**

146. The domain of the function $f(x) = \ln\left(x^2 - 4x + 4\right)$ is which of the
    following?
    a. $(2, \infty)$
    b. $(-\infty, 2)$
    c. $(-\infty, 2) \cup (2, \infty)$
    d. $\mathbb{R}$

147. Which of the following is a characteristic of the graph of
$f(x) = \ln(x+1)+1$?
  a. The y-intercept is $(e,1)$.
  b. $x = -1$ is a vertical asymptote.
  c. There is no x-intercept.
  d. $y = 1$ is a horizontal asymptote.

148. Which of the following are characteristics of the graph of
$f(x) = -e^{2-x} - 3$?
  a. The graph of f lies below the x-axis.
  b. $y = -3$ is the horizontal asymptote for the graph of f.
  c. The domain is $\mathbb{R}$.
  d. All of the above are characteristics of the graph.

149. True or false? If $f(x) = e^{2x}$ and $g(x) = \ln \sqrt{x}$, $x > 0$, then f and g are inverses.

150. Write the following as the logarithm of a single expression:
$3\ln\left(xy^2\right) - 4\ln\left(x^2 y\right) + \ln(xy)$

151. Determine the values of x that satisfy the equation $2^{7x^2-1} = 4^{3x}$.
  a. $x = \dfrac{-3 \pm \sqrt{37}}{14}$
  b. $x = \dfrac{3 \pm \sqrt{37}}{14}$
  c. $x = -\frac{1}{7}$ and $x = 1$
  d. $x = \frac{1}{7}$ and $x = -1$

152. Determine the values of x, if any, that satisfy the equation $5^{\sqrt{x+1}} = \frac{1}{25}$.
  a. 3
  b. −3
  c. $\log_5 3$
  d. There is no solution.

153. Determine the value of x that satisfies the equation
$\ln(x-2) - \ln(3-x) = 1$.

    a. $\frac{3e+2}{e+1}$

    b. $\frac{3(e+2)}{e+1}$

    c. 2

    d. 3

154. Determine the solution set for the inequality $5 \le 4e^{2-3x} + 1 < 9$.

    a. $\left(\frac{2-\ln 2}{3}, \frac{e-2}{3} \right]$

    b. $\left(\frac{2-\ln 2}{3}, \frac{2}{3} \right]$

    c. $\left[\frac{2}{3}, \frac{-2+\ln 2}{3} \right)$

    d. $\left[\frac{e-2}{3}, \frac{-2+\ln 2}{3} \right)$

155. Determine the solution set for the inequality $\ln(1-x^2) \le 0$.

    a. $(-1,0) \cup (0,1)$

    b. $(-\infty,-1) \cup (1,\infty)$

    c. $(-1,1)$

    d. $[-1,1]$

Answers

126. Finding x such that $\log_3 27 = x$ is equivalent to finding x such that $3^x = 27$. Since $27 = 3^3$, the solution of this equation is 3.

127. Finding x such that $\log_3\left(\frac{1}{9}\right) = x$ is equivalent to finding x such that $3^x = \frac{1}{9}$. Since $\frac{1}{9} = 3^{-2}$, we conclude that the solution of this equation is -2.

128. Finding x such that $\log_{\frac{1}{2}} 8 = x$ is equivalent to finding x such that $\left(\frac{1}{2}\right)^x = 8$. Since $8 = 2^3$ and $\left(\frac{1}{2}\right)^x = (2^{-1})^x = 2^{-x}$, this equation is equivalent to $2^3 = 2^{-x}$, the solution of which is -3.

129. Finding x such that $\log_7 \sqrt{7} = x$ is equivalent to finding x such that $7^x = \sqrt{7}$. Since $\sqrt{7} = 7^{\frac{1}{2}}$, we conclude that the solution of this equation is $\frac{1}{2}$.

130. Finding x such that $\log_5 1 = x$ is equivalent to finding x such that $5^x = 1$. We conclude that the solution of this equation is 0.

131. Finding x such that $\log_{16} 64 = x$ is equivalent to finding x such that $16^x = 64$. We rewrite the expressions on both sides of the equation using the same base, namely 2. Indeed, observe that $16^x = \left(2^4\right)^x = 2^{4x}$ and $64 = 2^6$. Hence, the value of x we seek is the solution of the equation $4x = 6$, namely $x = \frac{6}{4} = \frac{3}{2}$.

132. The equation $\log_6 x = 2$ is equivalent to $x = 6^2 = 36$. So, the solution is $x = 36$.

133. $\log_a x = \log_a\left(5\sqrt{a}\right) = \log_a 5 + \log_a\left(\sqrt{a}\right) = \log_a 5 + \log_a\left(a^{\frac{1}{2}}\right)$
$$= \log_a 5 + \frac{1}{2}\underbrace{\log_a a}_{=1} = \log_a 5 + \frac{1}{2}$$

134. $\log_3\left(3^4 \cdot 9^3\right) = \log_3\left(3^4\right) + \log_3\left(9^3\right) = 4\log_3 3 + 3\log_3 9 = 4(1) + 3(2) = 10$

135. The equation $5^{3x-1} = 7$ is equivalent to $\log_5 7 = 3x - 1$. This equation is solved as follows:

$$\log_5 7 = 3x - 1$$

$$3x = 1 + \log_5 7$$

$$x = \tfrac{1}{3}\left(1 + \log_5 7\right)$$

Note: You can further apply the change of base formula $\log_a x = \frac{\ln x}{\ln a}$ to further simply x as $x = \tfrac{1}{3}\left(1 + \log_5 7\right) = \tfrac{1}{3}\left(1 + \frac{\ln 7}{\ln 5}\right)$.

136. The given expression can be written as one involving the terms $\log_a x$ and $\log_a y$, as follows:

$$\log_a\left(\tfrac{x}{y^3}\right) = \log_a x - \log_a(y^3) = \log_a x - 3\log_a y$$

Substituting $\log_a x = 2$ and $\log_a y = -3$ into this expression yields

$$\log_a\left(\tfrac{x}{y^3}\right) = \log_a x - 3\log_a y = 2 - 3(-3) = 2 + 9 = 11$$

137. Since $f(x) = \log_a x$ and $g(x) = a^x$ are inverses, it follows by definition that $g(f(x)) = a^{\log_a x} = x$. Hence, $3^{\log_3 2}$ reduces to just 2, so $3^{\log_3 2} = 2$.

138. Since $f(x) = \log_a x$ and $g(x) = a^x$ are inverses, it follows by definition that $f(g(x)) = \log_a\left(a^x\right) = x$.

139. First, write the expression on the left side as the ln of a single expression. Then, use property 5 noted at the beginning of the chapter, as follows:

$$3\ln\left(\tfrac{1}{x}\right) = \ln 8$$

$$\ln\left(\tfrac{1}{x}\right)^3 = \ln 8$$

$$\left(\tfrac{1}{x}\right)^3 = 8$$

$$x^{-3} = 8$$

$$\left(x^{-3}\right)^{\frac{-1}{3}} = 8^{\frac{-1}{3}}$$

$$x = \tfrac{1}{2}$$

140. $e^{-\frac{1}{2}\ln 3} = e^{\ln\left(3^{-\frac{1}{2}}\right)} = 3^{-\frac{1}{2}} = \frac{1}{\sqrt{3}} = \frac{\sqrt{3}}{3}$

141. The given expression can be written as one involving the terms $\ln x$ and $\ln y$ as follows:

$$\ln\left(\frac{e^2 y}{\sqrt{x}}\right) = \ln\left(e^2 y\right) - \ln(\sqrt{x}) = \ln\left(e^2\right) + \ln y - \ln\left(x^{\frac{1}{2}}\right) = 2\ln(e) + \ln y - \tfrac{1}{2}\ln x$$

$$= 2 + \ln y - \tfrac{1}{2}\ln x$$

Substituting $\ln x = 3$ and $\ln y = 2$ into this expression yields

$$\ln\left(\frac{e^2 y}{\sqrt{x}}\right) = 2 + \ln y - \tfrac{1}{2}\ln x = 2 + 2 - \tfrac{1}{2}(3) = 4 - \tfrac{3}{2} = \tfrac{5}{2}$$

142. Applying the exponent rules enables us to rewrite the given expression as follows:

$$e^{3x-2y} = e^{3x} \cdot e^{-2y} = (e^x)^3 \cdot (e^y)^{-2}$$

Substituting $e^x = 2$ and $e^y = 3$ into this expression yields

$$e^{3x-2y} = e^{3x} \cdot e^{-2y} = (e^x)^3 \cdot (e^y)^{-2} = 2^3 \cdot 3^{-2} = \tfrac{8}{9}$$

143. b. The graph of $f(x) = 1 - 2e^x$ is obtained by reflecting the graph of $g(x) = e^x$ over the x-axis, scaling it by a factor of 2, and then translating it up one unit. In doing so, the original horizontal asymptote $y = 0$ for g becomes $y = 1$, and the graph of f always stays below this asymptote. Hence, the range is $(-\infty, 1)$. The graph is provided here.

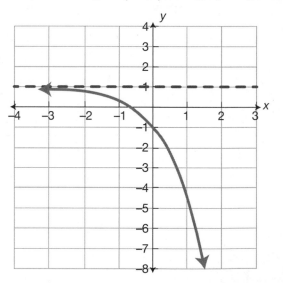

144. d. The range of the function $g(x) = \ln(2x - 1)$ is the set of all real numbers, $\mathbb{R}$.

Since using $2x - 1$ as the input for g for all $x > \frac{1}{2}$ covers the same set of inputs as g, the range of g is also the set of all real numbers, $\mathbb{R}$. The graph is provided here.

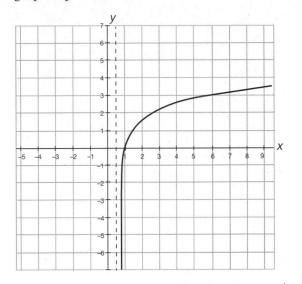

145. c. The x-intercepts of the function $f(x) = \ln\left(x^2 - 4x + 4\right)$ are those x-values that satisfy the equation $x^2 - 4x + 4 = 1$, solved as follows:

$x^2 - 4x + 4 = 1$
$x^2 - 4x + 3 = 0$
$(x - 3)(x - 1) = 0$

The solutions of this equation are $x = 3$ and $x = 1$. So, the x-intercepts are $(1,0)$ and $(3,0)$.

146. c. The domain of the function $f(x) = \ln\left(x^2 - 4x + 4\right)$ is the set of x-values for which $x^2 - 4x + 4 > 0$. This inequality is equivalent to $(x - 2)^2 > 0$, which is satisfied by all real numbers x except 2. So, the domain is $(-\infty, 2) \cup (2, \infty)$.

147. b. The vertical asymptote for $f(x) = \ln(x + 1) + 1$ occurs at those x-values that make the input of the ln portion equal to zero, namely $x = -1$.

148. d. The graph of $f(x) = -e^{2-x} - 3$ can be obtained by reflecting the graph of $g(x) = e^{-x}$ about the y-axis, then shifting it to the left two units, then reflecting it over the x-axis, and finally shifting it down three units. The graph is as follows:

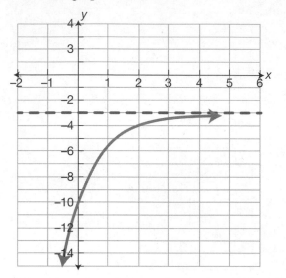

It is evident that all three characteristics provided in choices **a** through **c** hold.

149. True. Using exponent and logarithm rules yields

$$f(g(x)) = e^{2(\ln\sqrt{x})} = e^{2\ln\left(x^{\frac{1}{2}}\right)} = e^{2\left(\frac{1}{2}\right)\ln x} = e^{\ln x} = x \text{ , for all } x > 0$$

$$g(f(x)) = \ln\sqrt{e^{2x}} = \ln\sqrt{(e^x)^2} = \ln(e^x) = x, \text{ for all real numbers } x$$

Hence, we conclude that f and g are inverses.

150. Using the logarithm rules yields

$$3\ln(xy^2) - 4\ln(x^2 y) + \ln(xy) = \ln(xy^2)^3 - \ln(x^2 y)^4 + \ln(xy)$$

$$= \ln\left[\frac{(xy^2)^3}{(x^2 y)^4}\right] + \ln(xy) = \ln\left[\frac{(xy^2)^3 (xy)}{(x^2 y)^4}\right]$$

$$= \ln\left[\frac{x^3 y^6 xy}{x^8 y^4}\right] = \ln\left[\frac{y^3}{x^4}\right]$$

151. c. Rewrite the expression on the right side of the equation as a power of 2, as $4^{3x} = \left(2^2\right)^{3x} = 2^{6x}$, and substitute into the original equation to obtain the equivalent equation $2^{7x^2-1} = 2^{6x}$. Now, equate the exponents and solve for x, as follows:

$7x^2 - 1 = 6x$
$7x^2 - 6x - 1 = 0$
$(7x+1)(x-1) = 0$

The solutions are $x = -\frac{1}{7}$ and $x = 1$.

152. d. Since $\frac{1}{25} = 5^{-2}$, the original equation is equivalent to $5^{\sqrt{x+1}} = 5^{-2}$. The x-values that satisfy this equation must satisfy the one obtained by equating the exponents of the expressions on both sides of the equation, namely $\sqrt{x+1} = -2$. But the left side of this equation is nonnegative for any x-value that does not make the radicand negative. Hence, the equation has no solution.

153. a. Apply the logarithm properties and then solve for x, as follows:

$\ln(x-2) - \ln(3-x) = 1$

$\ln\left(\frac{x-2}{3-x}\right) = 1$

$\frac{x-2}{3-x} = e$

$x - 2 = e(3-x)$
$x - 2 = 3e - ex$
$ex + x = 3e + 2$
$(e+1)x = 3e + 2$

$x = \frac{3e+2}{e+1}$

Substituting this value back in for x in the original equation reveals that it is indeed a solution.

154. b.

$5 \le 4e^{2-3x} + 1 < 9$
$4 \le 4e^{2-3x} < 8$
$1 \le e^{2-3x} < 2$
$\ln 1 \le 2 - 3x < \ln 2$
$-2 + 0 \le -3x < -2 + \ln 2$

$\frac{2}{3} \ge x > \frac{-2+\ln 2}{-3}$

$\frac{2-\ln 2}{3} < x \le \frac{2}{3}$

So, the solution set is $\left(\frac{2-\ln 2}{3}, \frac{2}{3}\right]$.

155. **c.** First, note that the expression $\ln\left(1-x^2\right)$ is defined only when $1-x^2 > 0$, which is equivalent to $-1 < x < 1$. Since the only values of y for which $\ln y \le 0$ are $0 < y \le 1$, we must determine which of these values satisfies the more restrictive inequality $0 < 1-x^2 \le 1$. But notice that it is true for every x in the interval $(-1, 1)$. Hence, the solution set of the inequality $\ln\left(1-x^2\right) \le 0$ is $(-1, 1)$.

7

Trigonometry Problems

The standard notions of sine, cosine, tangent, cotangent, secant, and cosecant that one encounters in elementary triangle trigonometry can be extended to the setting of real-valued functions using the so-called *unit circle* approach.

The circle of radius 1 around the origin is called the *unit circle*. As such, the hypotenuse has length 1, and the sine is the y-value of the point where a ray of the given angle intersects with the circle of radius 1. Similarly, the cosine is the x-value. Note in the following figure that the angle of measure 0 runs straight to the right along the positive x-axis, and every other positive angle is measured counterclockwise from there.

This can be used to find the trigonometric values of nice angles greater than $90° = \frac{\pi}{2}$. The trick is to use either a 30°, 60°, 90° triangle (a $\frac{\pi}{6}$, $\frac{\pi}{3}$, $\frac{\pi}{2}$ triangle) or else a 45°, 45°, 90° triangle (a $\frac{\pi}{4}$, $\frac{\pi}{4}$, $\frac{\pi}{2}$ triangle) to find the y-value (sine) and x-value (cosine) of the appropriate point on the unit circle.

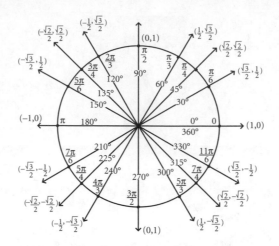

Properties of Trigonometric Functions

The following are some basic properties of trigonometric functions, which hold for all x-values for which the expressions are defined.

1. $\sin(-x) = -\sin x$
2. $\cos(-x) = \cos x$
3. $\tan x = \frac{\sin x}{\cos x}$
4. $\cot x = \frac{1}{\tan x} = \frac{\cos x}{\sin x}$
5. $\sec x = \frac{1}{\cos x}$
6. $\csc x = \frac{1}{\sin x}$
7. $\sin(x + 2n\pi) = \sin x$, for any integer n
8. $\cos(x + 2n\pi) = \cos x$, for any integer n
9. $\tan(x + n\pi) = \tan x$, for any integer n
10. $\sin^2 x + \cos^2 x = 1$
11. $1 + \tan^2 x = \sec^2 x$
12. $\sin 2x = 2\sin x \cos x$
13. $\cos 2x = \cos^2 x - \sin^2 x = 2\cos^2 x - 1 = 1 - 2\sin^2 x$
14. $\sin(x \pm y) = \sin x \cos y \pm \sin y \cos x$
15. $\cos(x \pm y) = \cos x \cos y \mp \sin x \sin y$

The following review of the basics of unit circle trigonometry includes questions on the graphs of these functions and translations and reflections of them.

Questions

For Questions 156 through 160, simplify the given expression.

156. $\sin\left(\frac{\pi}{4}\right) = $ _____

157. $\tan\left(\frac{5\pi}{6}\right) = $ _____

158. $\csc\left(-\frac{2\pi}{3}\right) = $ _____

159. $\sec(7\pi) = $ _____

160. $\cot\left(-\frac{113\pi}{2}\right) = $ _____

For Questions 161 through 165, write each expression in terms of $\sin\alpha$, $\cos\alpha$, $\sin\beta$, $\cos\beta$, and real numbers.

161. $\sin^2\alpha + \cos^2\alpha = $ _____

162. $\sin 2\alpha = $ _____

163. $\cos 2\beta = $ _____

164. $\sin(\alpha - \beta) = $ _____

165. $1 + \tan^2\alpha = $ _____

166. If $\csc\alpha = \frac{5}{3}$ and $\cos\beta = \frac{8}{17}$ and the terminal sides of both angles lie in the first quadrant, then $\sin(\alpha + \beta) = $ _____.

167. If $\cos\beta = \frac{8}{17}$ and the terminal side of β lies in the fourth quadrant, then $\sin 2\beta = $ _____.

168. Determine the values of x in $[0, 2\pi)$ that satisfy $\sin x + 2\cos^2 x = 1$.

169. Sketch the graph of $f(x) = 2\sin x - 1$ on $[0, 4\pi]$.

170. Which of the following are the x-intercepts of $f(x) = \tan\left(2x + \frac{\pi}{6}\right)$?

    a. $\left(\frac{5n\pi}{2}, 0\right)$, where n is an integer

    b. $\left(\frac{(6n-1)\pi}{12}, 0\right)$, where n is an integer

    c. $\left(\frac{(3n+1)\pi}{6}, 0\right)$, where n is an integer

    d. There are no x-intercepts.

171. Which of the following is the y-intercept of $f(x) = -3\cos\left(4x + \frac{\pi}{3}\right) - \frac{1}{2}$?

    a. $(0, -2)$

    b. $\left(0, \frac{-3\sqrt{2}-1}{2}\right)$

    c. $\left(0, -\frac{1}{2}\right)$

    d. $(0, 2)$

172. Which of the following expressions is equivalent to $\sin 3x$?

    a. $\cos\left(3x - \frac{\pi}{2}\right)$

    b. $\sin x \cos 2x + \sin 2x \cos x$

    c. $2\sin\left(\frac{3}{2}x\right)\cos\left(-\frac{3}{2}x\right)$

    d. all of the above

173. Determine the x-values in the interval $[0, 2\pi]$ that satisfy the inequality $0 < \sin x \leq \frac{\sqrt{2}}{2}$.

    a. $\left(0, \frac{\pi}{4}\right] \cup \left[\frac{3\pi}{4}, \pi\right)$

    b. $\left(0, \frac{\pi}{6}\right] \cup \left[\frac{5\pi}{6}, \pi\right)$

    c. $\left(\frac{\pi}{6}, \frac{5\pi}{6}\right]$

    d. $\left[\frac{\pi}{4}, \frac{3\pi}{4}\right)$

174. Simplify the expression $\sin\left(-\frac{\pi}{4}\right) + 3\cos\left(\frac{7\pi}{6}\right) - \tan(\pi)$.

175. Which of the following expressions is equivalent to $\frac{\sin^2 x + \cos^2 x}{\cos(-x)}$?

    a. $-\sec x$

    b. $-\csc x$

    c. $\sec x$

    d. $\csc x$

176. Which of the following expressions is equivalent to $\frac{\cot x \sec x}{\tan x \csc x}$?

    a. $\cot x$

    b. $\tan x$

    c. $\sec x$

    d. $\csc x$

177. True or false? The domain of $f(x) = \tan 2x$ is the set of all real numbers EXCEPT all odd multiples of $\frac{\pi}{4}$.

178. True or false? The vertical asymptotes of the graph of $f(x) = \cot(x - \pi)$ are the lines $y = n\pi$, where n is an integer.

179. Which of the following is the maximum y-value attained by the graph of $f(x) = -4\sin(\pi x) + 1$?

    a. 1

    b. 3

    c. 4

    d. 5

180. Determine the x-values that satisfy the equation $\cos\left(x - \frac{\pi}{3}\right) = -\frac{1}{2}$.

181. Determine all points of intersection of the graphs of $y = \sin x$ and $y = \cos x$.

182. Which of the following is the range of the function $f(x) = 5\cos\left(2x - \frac{\pi}{8}\right) + 3$?

    a. $[1,3]$

    b. $[-2,8]$

    c. $[-5,5]$

    d. $\mathbb{R}$

183. Which of the following is the period of the graph of
$f(x) = 5\cos\left(2x - \frac{\pi}{8}\right) + 3$?

a. $\frac{2\pi}{3}$

b. 16

c. $\frac{2\pi}{5}$

d. π

184. What is the minimum y-value attained by the graph of $y = -\tan 2x$?

a. 0

b. $\frac{\pi}{2}$

c. -1

d. No such minimum value exists.

185. Determine the x-values in the interval $\left[-\frac{\pi}{2}, \frac{3\pi}{2}\right]$ that satisfy the inequality $\tan x > -1$.

Answers

156. Appealing to the unit circle depicting the sine and cosine of the standard angles in the interval $[0, 2\pi]$, we conclude that $\sin\left(\frac{\pi}{4}\right) = \frac{\sqrt{2}}{2}$.

157. Appealing to the unit circle depicting the sine and cosine of the standard angles in $[0, 2\pi]$, as well as the definition of tangent, we see that

$$\tan\left(\frac{5\pi}{6}\right) = \frac{\sin\left(\frac{5\pi}{6}\right)}{\cos\left(\frac{5\pi}{6}\right)} = \frac{\frac{1}{2}}{-\frac{\sqrt{3}}{2}} = -\frac{1}{\sqrt{3}} = -\frac{\sqrt{3}}{3}$$

158. Appealing to the unit circle depicting the sine and cosine of the standard angles in $[0, 2\pi]$, as well as the definition of cosecant, we see that

$$\csc\left(-\frac{2\pi}{3}\right) = \frac{1}{\sin\left(-\frac{2\pi}{3}\right)} = \frac{1}{-\sin\left(\frac{2\pi}{3}\right)} = -\frac{1}{\frac{\sqrt{3}}{2}} = -\frac{2}{\sqrt{3}} = -\frac{2\sqrt{3}}{3}$$

159. Appealing to the unit circle depicting the sine and cosine of the standard angles in $[0, 2\pi]$, the definition of secant, and the periodicity of cosine, we see that

$$\sec(7\pi) = \frac{1}{\cos(7\pi)} = \frac{1}{\cos(\pi + 6\pi)} = \frac{1}{\cos(\pi)} = \frac{1}{-1} = -1$$

160. Appealing to the unit circle depicting the sine and cosine of the standard angles in $[0, 2\pi]$, the definition of cotangent, and the periodicity of cosine and sine, we see that

$$\cot\left(-\frac{113\pi}{2}\right) = \frac{\cos\left(-\frac{113\pi}{2}\right)}{\sin\left(-\frac{113\pi}{2}\right)} = \frac{\cos\left(\frac{113\pi}{2}\right)}{-\sin\left(\frac{113\pi}{2}\right)} = \frac{\cos\left(\frac{\pi}{2} + 2(28)\pi\right)}{-\sin\left(\frac{\pi}{2} + 2(28)\pi\right)} = \frac{\cos\left(\frac{\pi}{2}\right)}{-\sin\left(\frac{\pi}{2}\right)} = -\frac{0}{1} = 0$$

161. Using the properties mentioned at the beginning of the chapter, we conclude that $\sin^2 \alpha + \cos^2 \alpha = 1$.

162. Using the properties mentioned at the beginning of the chapter, we conclude that $\sin 2\alpha = 2\sin\alpha\cos\alpha$.

163. There are several equivalent answers here. Using the properties mentioned at the beginning of the chapter, we conclude that
$\cos 2\beta = \cos^2 \beta - \sin^2 \beta$
Equivalently, using the fact that $\cos^2 \beta + \sin^2 \beta = 1$, we could solve this equation for either $\cos^2 \beta$ or $\sin^2 \beta$ to obtain $\cos^2 \beta = 1 - \sin^2 \beta$ and $\sin^2 \beta = 1 - \cos^2 \beta$, respectively. Substituting these into the preceding formula yields the following two equivalent formulas:
$\cos 2\beta = \cos^2 \beta - \sin^2 \beta = (1 - \sin^2 \beta) - \sin^2 \beta = 1 - 2\sin^2 \beta$
$\cos 2\beta = \cos^2 \beta - \sin^2 \beta = \cos^2 \beta - (1 - \cos^2 \beta) = 2\cos^2 \beta - 1$

164. Using the properties mentioned at the beginning of the chapter, we conclude that $\sin(\alpha - \beta) = \sin\alpha\cos\beta - \sin\beta\cos\alpha$.

165. Using the properties mentioned at the beginning of the chapter, we conclude that $1 + \tan^2 \alpha = \sec^2 \alpha$.

166. Using the properties mentioned at the beginning of the chapter, we see that $\sin(\alpha + \beta) = \sin\alpha\cos\beta + \sin\beta\cos\alpha$.

We are given that $\csc\alpha = \frac{5}{3}$ and $\cos\beta = \frac{8}{17}$. By definition of cosecant, $\csc\alpha = \frac{5}{3}$ is equivalent to $\sin\alpha = \frac{3}{5}$. Also, since the terminal sides of both α and β are stated to lie in the first quadrant, it follows that both $\sin\beta$ and $\cos\alpha$ are positive. Hence, using the identity $\sin^2 x + \cos^2 x = 1$, we see that

$\sin^2 \alpha + \cos^2 \alpha = 1 \quad \sin^2 \beta + \cos^2 \beta = 1$

$\left(\frac{3}{5}\right)^2 + \cos^2 \alpha = 1 \quad \sin^2 \beta + \left(\frac{8}{17}\right)^2 = 1$

$\cos^2 \alpha = \frac{16}{25} \quad \sin^2 \beta = \frac{225}{289}$

$\cos\alpha = \frac{4}{5} \quad \sin\beta = \frac{15}{17}$

Substituting these values into the initial equation yields

$\sin(\alpha + \beta) = \sin\alpha\cos\beta + \sin\beta\cos\alpha = \left(\frac{3}{5}\right)\left(\frac{8}{17}\right) + \left(\frac{15}{17}\right)\left(\frac{4}{5}\right) = \frac{84}{85}$.

167. Using the properties mentioned at the beginning of the chapter, we see that

$\sin 2\beta = 2\sin\beta\cos\beta$.

Since the terminal side of β is assumed to lie in the fourth quadrant, it follows that $\sin\beta$ is negative. Hence, using the identity $\sin^2\beta + \cos^2\beta = 1$, we see that

$\sin^2\beta + \cos^2\beta = 1$

$\sin^2\beta + \left(\frac{8}{17}\right)^2 = 1$

$\sin^2\beta = \frac{225}{289}$

$\sin\beta = -\frac{15}{17}$

Substituting these values into the initial equation yields

$\sin 2\beta = 2\sin\beta\cos\beta = 2\left(\frac{8}{17}\right)\left(-\frac{15}{17}\right) = -\frac{240}{289}$

168. Express $\cos^2 x$ in terms of $\sin x$ using the identity $\sin^2 x + \cos^2 x = 1$, and simplify, as follows:

$\sin x + 2\left(1 - \sin^2 x\right) = 1$

$-2\sin^2 x + \sin x + 2 = 1$

$2\sin^2 x - \sin x - 1 = 0$

$(2\sin x + 1)(\sin x - 1) = 0$

Now, we set each factor equal to zero and determine the values of x in $[0, 2\pi)$ that satisfy each of them, as follows:

$2\sin x + 1 = 0$ $\qquad$ $\sin x - 1 = 0$

$\sin x = -\frac{1}{2}$ $\qquad$ $\sin x = 1$

$x = \frac{7\pi}{6}, \frac{11\pi}{6}$ $\qquad$ $x = \frac{\pi}{2}$

Hence, the solutions of the original equation are $x = \frac{\pi}{2}, \frac{7\pi}{6}$, and $\frac{11\pi}{6}$.

169. We first graph $y = 2\sin x$, which has the same x-intercepts as $y = \sin x$, at all integer multiples of π, but has maximum and minimum values of 2 and -2, respectively. This graph is:

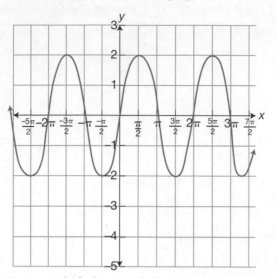

Now, we shift this graph down one unit to obtain the graph of $f(x) = 2\sin x - 1$, as follows:

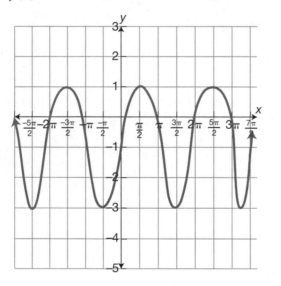

170. b. By definition of tangent, we see that

$$\tan\left(2x + \tfrac{\pi}{6}\right) = \frac{\sin\left(2x + \tfrac{\pi}{6}\right)}{\cos\left(2x + \tfrac{\pi}{6}\right)}$$

The x-intercepts are those values for which $f(x) = 0$, and this occurs if and only if $\sin\left(2x + \tfrac{\pi}{6}\right) = 0$. This equation is satisfied by those x-values which satisfy the equation $2x + \tfrac{\pi}{6} = n\pi$, where n is an integer, which is solved as follows:

$$2x + \tfrac{\pi}{6} = n\pi$$

$$2x = n\pi - \tfrac{\pi}{6} = \tfrac{6n\pi - \pi}{6} = \tfrac{(6n-1)\pi}{6}$$

$$x = \tfrac{(6n-1)\pi}{12}$$

Thus, the x-intercepts of f are $\left(\tfrac{(6n-1)\pi}{12}, 0\right)$, where n is an integer.

171. a. The y-intercept of $f(x) = -3\cos\left(4x + \tfrac{\pi}{3}\right) - \tfrac{1}{2}$ is the point $(0, f(0))$. Observe that

$$f(0) = -3\cos\left(4(0) + \tfrac{\pi}{3}\right) - \tfrac{1}{2} = -3\cos\left(\tfrac{\pi}{3}\right) - \tfrac{1}{2} = -3\left(\tfrac{1}{2}\right) - \tfrac{1}{2} = -2$$

Thus, the y-intercept is $(0, -2)$.

172. d. Each of these expressions is equivalent to $\sin 3x$, and each is obtained by applying a different identity. Specifically, using properties 12, 14, and 15 provided at the beginning of the chapter yields the following:

$$\cos\left(3x - \tfrac{\pi}{2}\right) = \cos(3x)\cos\left(\tfrac{\pi}{2}\right) + \sin(3x)\sin\left(\tfrac{\pi}{2}\right) = \cos(3x)(0) + \sin(3x)(1)$$
$$= \sin 3x$$

$$\sin x \cos 2x + \sin 2x \cos x = \sin(x + 2x) = \sin 3x$$

$$2\sin\left(\tfrac{3}{2}x\right)\cos\left(-\tfrac{3}{2}x\right) = 2\sin\left(\tfrac{3}{2}x\right)\cos\left(\tfrac{3}{2}x\right) = \sin\left(2 \cdot \left(\tfrac{3}{2}x\right)\right) = \sin 3x$$

173. a. First, the x-values in $[0, 2\pi]$ for which $\sin x > 0$ are those that lie in $(0, \pi)$. Of these, the only x-values satisfying $0 < \sin x \le \tfrac{\sqrt{2}}{2}$ are those in the set $\left(0, \tfrac{\pi}{4}\right] \cup \left[\tfrac{3\pi}{4}, \pi\right)$.

174. Appealing to the unit circle depicting the sine and cosine of the standard angles in $[0, 2\pi]$ and using the identity $\sin(-x) = -\sin x$ yields

$$\sin\left(-\tfrac{\pi}{4}\right) = -\sin\left(\tfrac{\pi}{4}\right) = -\tfrac{\sqrt{2}}{2}$$

$$\cos\left(\tfrac{7\pi}{6}\right) = -\tfrac{\sqrt{3}}{2}$$

$$\tan(\pi) = 0$$

Substituting these into the given expression yields

$$\sin\left(-\tfrac{\pi}{4}\right) + 3\cos\left(\tfrac{7\pi}{6}\right) - \tan(\pi) = -\tfrac{\sqrt{2}}{2} - \tfrac{3\sqrt{3}}{2} - 0 = -\tfrac{\sqrt{2} + 3\sqrt{3}}{2}$$

175. c. Using identities 10, then 2, and then 5 provided at the beginning of the chapter yields

$$\frac{\sin^2 x + \cos^2 x}{\cos(-x)} = \frac{1}{\cos(-x)} = \frac{1}{\cos x} = \sec x$$

176. a. Using identities 3 through 6 provided at the beginning of the chapter yields

$$\frac{\cot x \sec x}{\tan x \csc x} = \frac{\frac{\cos x}{\sin x} \cdot \frac{1}{\cos x}}{\frac{\sin x}{\cos x} \cdot \frac{1}{\sin x}} = \frac{\frac{1}{\sin x}}{\frac{1}{\cos x}} = \frac{\cos x}{\sin x} = \cot x$$

177. True. Using the definition of tangent, we see that

$$f(x) = \tan 2x = \frac{\sin 2x}{\cos 2x}$$

The domain of $f(x) = \tan 2x$ is the set of all real numbers x for which $\cos 2x \neq 0$. To determine the precise values of x that are excluded from the domain, we solve $\cos 2x = 0$. The x-values that satisfy this equation are those for which $2x = \frac{(2n+1)\pi}{2}$, or equivalently $x = \frac{(2n+1)\pi}{4}$, where n is an integer. Verbally, the values excluded from the domain of f are the odd multiples of $\frac{\pi}{4}$.

178. False. Using the definition of cotangent, we see that

$$f(x) = \cot(x - \pi) = \frac{\cos(x - \pi)}{\sin(x - \pi)}$$

The vertical asymptotes of the graph of $f(x) = \cot(x - \pi)$ occur at precisely those x-values that make the denominator, $\sin(x - \pi)$, equal to zero. These values satisfy the equation $x - \pi = n\pi$, or equivalently $x = (n+1)\pi$, where n is an integer. The lines described in the original statement, those with equation $y = n\pi$, where n is an integer, are horizontal, not vertical.

179. d. Begin by graphing $y = 4\sin(\pi x)$ and reflecting it over the x-axis. The maximum value that this graph attains is 4. Now, shift this graph up one unit to attain the graph of f. Since 1 is being added to each y-value of the previous graph, we conclude that the maximum y-value attained by the graph of f is 5.

180. The equation $\cos\left(x - \frac{\pi}{3}\right) = -\frac{1}{2}$ is satisfied precisely when x satisfies one of the following two equations:

$x - \frac{\pi}{3} = \frac{2\pi}{3} + 2n\pi$, where n is an integer

$x - \frac{\pi}{3} = \frac{4\pi}{3} + 2n\pi$, where n is an integer

The solutions of these equations are:

$x = \pi + 2n\pi = (2n+1)\pi$, where n is an integer

$x = \frac{5\pi}{3} + 2n\pi = \frac{5\pi + 6n\pi}{3} = \frac{(5+6n)\pi}{3}$, where n is an integer

181. The x-values of the points of intersection of the graphs of $y = \sin x$ and $y = \cos x$ satisfy the equation $\sin x = \cos x$. The x-values in $[0, 2\pi]$ that satisfy this equation are $x = \frac{\pi}{4}$ and $\frac{5\pi}{4}$. The coordinates of the points of intersection are thus $\left(\frac{\pi}{4}, \frac{\sqrt{2}}{2}\right)$ and $\left(\frac{5\pi}{4}, -\frac{\sqrt{2}}{2}\right)$. The graphs are provided here.

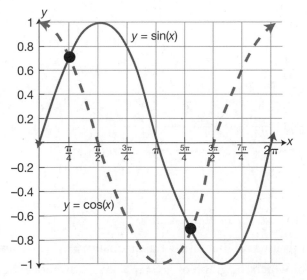

182. b. We must determine the minimum and maximum y-values attained by the graph of the function $f(x) = 5\cos\left(2x - \frac{\pi}{8}\right) + 3$. The graph of f is attained by shifting the graph of $y = 5\cos\left(2x - \frac{\pi}{8}\right)$ up three units. Since the minimum and maximum of the aforementioned graph are -5 and 5, respectively, adding 3 to each yields the lower and upper bounds of the range as being -2 and 8. Hence, the range of f is $[-2, 8]$.

183. d. The period of the graph of $y = A\cos(Bx + C) + D$, where A, B, C, and D are real numbers, is given by $\frac{2\pi}{B}$. The period of f is therefore $\frac{2\pi}{2} = \pi$.

184. d. The graph of $y = -\tan 2x$ has vertical asymptotes at all odd multiples of $\frac{\pi}{4}$, since such x-values make $\cos 2x = 0$, and hence make $\tan 2x$ undefined. Since the graph of $y = -\tan 2x$ follows each of these asymptotes to positive infinity from one side and negative infinity to the other, the graph never attains a minimum y-value.

185. Compare the graphs of $y = \tan x$ (solid lines) and $y = -1$ (dashed line) on the interval $\left[-\frac{\pi}{2}, \frac{3\pi}{2}\right]$, provided here:

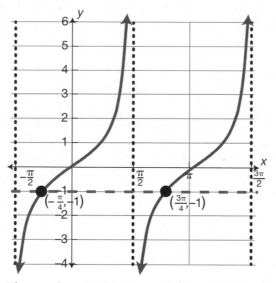

The x-values in this interval that satisfy the inequality $\tan x > -1$ are precisely those for which the graph of $y = \tan x$ is strictly above the graph of $y = -1$, namely $\left(-\frac{\pi}{4}, \frac{\pi}{2}\right) \cup \left(-\frac{3\pi}{4}, \frac{3\pi}{2}\right)$.

Limit Problems

The notion of a *limit* is the single most important underlying concept on which the calculus is built. We can use the notion of a limit to describe the behavior of a function *near* a particular input when it is not defined at the input. We first review some basic definitions and properties of limits.

Intuitive Definition: Let f be a function that is defined in the vicinity of a, but not necessarily at a. If it is the case that as $\underbrace{0 < |x - a| \to 0}_{\substack{\text{distance between } x \text{ and } a \\ \text{gets smaller from both sides}}}$, there exists a

corresponding real number L such that $\underbrace{0 < |f(x) - L| \to 0}_{\substack{\text{corresponding functional} \\ \text{values approach } L}}$, then we say f has

limit L at a and write $\lim\limits_{x \to a} f(x) = L$.

Basic Principles of Limits

Principle 1: A function need not be defined at $x = a$ in order to have a limit at $x = a$.

Principle 2: The functional value at $x = a$ is not relevant to the limit at $x = a$.

Principle 3: Knowing the functional value at $x = a$ is *not* sufficient to describe the function's behavior in a vicinity of $x = a$.

Principle 4: When a function oscillates too wildly near $x = a$, then there is no limit at $x = a$. (More precisely, if there are two sequences of inputs that approach $x = a$ for which the corresponding functional values approach *different* real numbers, then there is no limit at $x = a$.)

Principle 5: If the functional values approach L as x approaches a from the left, but the functional values approach M ($\neq L$) as x approaches a from the right, then there is no limit at $x = a$.

Principle 5 leads to the following definition.

Definition: f has a left-hand limit L at $x = a$ if as x approaches a from the left, the corresponding functional values approach the real number L. We write $\lim\limits_{x \to a^-} f(x) = L$. (The notion of a right-hand limit is defined in a similar manner.)

The notion of one-sided limits leads to the following useful characterization of limits.

Theorem

- If $\lim\limits_{x \to a^-} f(x) \neq \lim\limits_{x \to a^+} f(x)$ (or at least one of them does not exist), then $\lim\limits_{x \to a} f(x)$ does not exist.

- If $\lim\limits_{x \to a^-} f(x) = \lim\limits_{x \to a^+} f(x) = L$, then $\lim\limits_{x \to a} f(x) = L$.

The following rules enable us to compute the limits of various arithmetic combinations of functions.

Arithmetic of Limits

Let n and K be real numbers, and assume f and g are functions that have a limit at c.

Then, we have:

Rule (Symbolically)

1. $\lim_{x \to c} K = K$

2. $\lim_{x \to c} x = c$

3. $\lim_{x \to c} Kf(x) = K \lim_{x \to c} f(x)$

4. $\lim_{x \to c}\left[f(x) \pm g(x)\right] = \lim_{x \to c} f(x) \pm \lim_{x \to c} g(x)$

5. $\lim_{x \to c}\left[f(x)g(x)\right] = \left(\lim_{x \to c} f(x)\right) \cdot \left(\lim_{x \to c} g(x)\right)$

6. $\lim_{x \to c}\left[\dfrac{f(x)}{g(x)}\right] = \dfrac{\lim_{x \to c} f(x)}{\lim_{x \to c} g(x)}$, provided $\lim_{x \to c} g(x) \neq 0$

7. $\lim_{x \to c}\left[f(x)\right]^n = \left[\lim_{x \to c} f(x)\right]^n$, provided the latter is defined.

Rule (in Words)

1. Limit of a constant is the constant.
2. Limit of "x" as x goes to c is c.
3. Limit of a constant times a function is the constant times the limit of the function.
4. Limit of a sum (or difference) is the sum (or difference) of the limits.
5. Limit of a product is the product of the limits.
6. Limit of a quotient is the quotient of the limits, provided the denominator doesn't go to zero.
7. Limit of a function to a power is the power of the limit of the function.

All of these rules hold for left- and right-hand limits as well. If, upon applying these properties, the result is $\frac{0}{0}$ or $\frac{\infty}{\infty}$, you cannot apply them directly. Rather, some algebraic simplification must first occur. Then, reapply them.

Definition: A function $f(x)$ is continuous at $x = a$ if $\lim_{x \to a} f(x) = f(a)$.

Definition: $f(x)$ approaches negative (or positive) infinity as $x \to a$ if the corresponding functional values become unboundedly negative (or positive) as x approaches a.

We write $\lim\limits_{x \to a} f(x) = -\infty$ (or ∞).

We also can interpret such limits in terms of left- and right-hand limits. In all such cases, we say f is *unbounded* and call $x = a$ a *vertical asymptote* of f.

Definition: f has a limit L as x approaches ∞ (or $-\infty$) if the functional values can be made arbitrarily close to a single real number L for a sufficiently large x. We write $\lim\limits_{x \to \infty} f(x) = L$ or $\lim\limits_{x \to -\infty} f(x) = L$ and say the line $y = L$ is a *horizontal asymptote* of f.

Questions

For Questions 186 through 192, use the following graph to evaluate the given quantity.

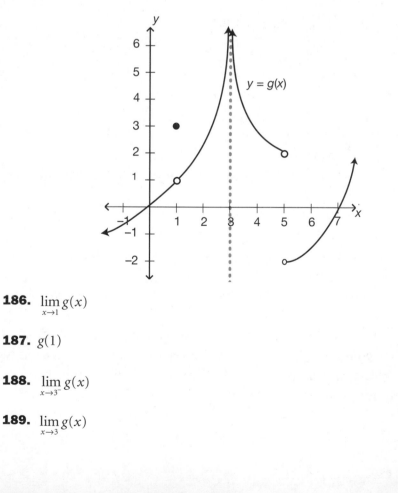

186. $\lim\limits_{x \to 1} g(x)$

187. $g(1)$

188. $\lim\limits_{x \to 3^-} g(x)$

189. $\lim\limits_{x \to 3} g(x)$

190. $\lim\limits_{x \to 5^-} g(x)$

191. $\lim\limits_{x \to 5^+} g(x)$

192. At which x-values in the interval $[-1,7]$ is g discontinuous?

For Questions 193 through 205, compute the indicated limit.

193. $\lim\limits_{x \to 4} \dfrac{4-x}{x^2+2x}$

194. $\lim\limits_{x \to \pi} \dfrac{3x}{(x-\pi)^3}$

195. $\lim\limits_{x \to -7} \dfrac{(3-x)(7+x)}{(x+7)(x+2)}$

196. $\lim\limits_{x \to 6^+} \dfrac{x+2}{6-x}$

197. $\lim\limits_{h \to 0} \dfrac{2(x+h)^3 - 2x^3}{h}$

198. $\lim\limits_{x \to \pi} \dfrac{\cos x + 1}{\cos x - 1}$
  a. 0
  b. The limit does not exist.
  c. -1
  d. -3

199. $\lim\limits_{x \to 3} \dfrac{x^3 + 2x^2 - 15x}{x^2 - 9}$
  a. 1
  b. -1
  c. 4
  d. The limit does not exist.

200. $\lim\limits_{x \to \frac{1}{4}} \dfrac{1 - 2\sqrt{x}}{1 - 4x}$
  a. $\dfrac{1}{2}$
  b. 2
  c. The limit does not exist.
  d. 1

201. $\lim\limits_{x\to 0^-} \frac{\sin x \cos x}{3x}$

   a. 0

   b. $\frac{1}{3}$

   c. $-\frac{1}{3}$

   d. The limit does not exist.

202. $\lim\limits_{x\to 0^+} \frac{1-e^x}{1-e^{2x}}$

   a. The limit does not exist.

   b. 1

   c. 2

   d. $\frac{1}{2}$

203. $\lim\limits_{x\to\left(-\frac{3\pi}{4}\right)^+} \tan\left(x+\frac{\pi}{4}\right)$

   a. $-\infty$

   b. ∞

   c. 0

   d. 1

204. $\lim\limits_{x\to(2e)^+} \ln(x-2e)$

   a. 0

   b. 1

   c. $-\infty$

   d. ∞

205. $\lim\limits_{x\to 4} \frac{4-x}{\frac{1}{x-5}+1}$

   a. 1

   b. −1

   c. The limit does not exist.

   d. 0

For Questions 206 through 215, use the following graphs to compute the indicated limit or answer the question posed.

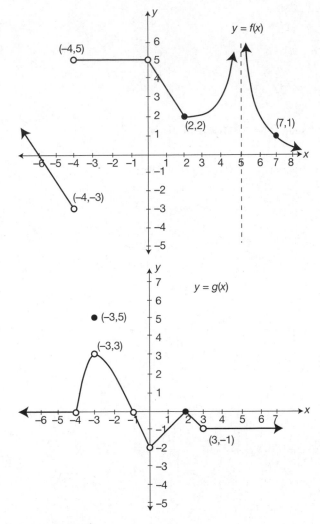

206. $\lim\limits_{x \to -4^+} \big(2g(x) - 3f(x)\big)$

207. $\lim\limits_{x \to 0} \big(f(x) \cdot g(x)\big)$

208. $\lim\limits_{x \to 5} \dfrac{g(x)}{f(x)}$

209. $\lim\limits_{x \to -3} -2\sqrt{g(x)}$

210. $\displaystyle\lim_{x\to 2}\frac{1}{g(x)}$

211. $\displaystyle\lim_{x\to -6^-}\frac{1}{f(x)}$

212. $\displaystyle\lim_{x\to 3}\sqrt[3]{8\,g(x)}$

213. $\displaystyle\lim_{x\to \infty}\bigl(g(x)+2f(x)\bigr)$

214. True or false? The function $h(x)=\dfrac{[f(x)]^3}{2g(x)+1}$ is continuous at $x=2$.

215. True or false? The function f has a limit at $x=-4$.

Answers

186. The y-values become closer to 1 as the x-values at which $g(x)$ is evaluated are taken closer to 1 from both the left and the right. Hence, $\lim_{x \to 1} g(x) = 1$. A common error is to say the limit is the functional value at 1, namely 3. Keep in mind that if you literally trace along the graph of $y = g(x)$ as x gets closer to 1 from both sides, the y-values on the graph get closer to 1, not 3.

187. The closed hole on the graph of g at the point $(1,3)$ implies that $g(1) = 3$.

188. The y-values become unboundedly large in the positive direction as the x-values at which $g(x)$ is evaluated are taken closer to 3 from the left. Hence, $\lim_{x \to 3^-} g(x) = \infty$.

189. The y-values become unboundedly large in the positive direction as the x-values at which $g(x)$ is evaluated are taken closer to 3 from both the left and the right. Hence, $\lim_{x \to 3} g(x) = \infty$.

190. The y-values become closer to 2 as the x-values at which $g(x)$ is evaluated are taken closer to 5 from the left. Hence, $\lim_{x \to 5^-} g(x) = 2$.

191. The y-values become closer to -2 as the x-values at which $g(x)$ is evaluated are taken closer to 5 from the right. Hence, $\lim_{x \to 5^+} g(x) = -2$.

192. The x-values in the interval $[-1,7]$ at which g is discontinuous are precisely those x-values at which any of the following occur: g has a limit at $x = a$ that is not equal to $g(a)$, g has a vertical asymptote at $x = a$, either the left-limit at $x = a$ or the right-limit at $x = a$ does not exist, or the left limit at $x = a$ is different from the right limit at $x = a$. Therefore, we conclude that g is discontinuous in this interval at 1, 3, and 5.

193. Substituting $x = 4$ directly into the expression yields $\frac{0}{4^2 + 2(4)} = 0$. Thus, we conclude that $\lim_{x \to 4} \frac{4 - x}{x^2 + 2x} = 0$.

194. Substituting $x = \pi$ directly into the expression yields $\frac{3\pi}{0}$, which suggests the presence of a vertical asymptote for the function $f(x) = \frac{3x}{(x - \pi)^3}$ at $x = \pi$. Note that substituting values close to π from the left side into this expression produces y-values that become unboundedly large in the negative direction, while substituting values close to π from the right side into this expression produces y-values that become unboundedly large in the positive direction. Hence, we conclude that $\lim_{x \to \pi} \frac{3x}{(x - \pi)^3}$ does not exist.

195. To compute this limit, cancel factors common to both the numerator and the denominator, and then substitute $x = -7$ into the simplified expression, as follows:

$$\lim_{x \to -7} \frac{(3-x)\cancel{(7+x)}}{\cancel{(x+7)}(x+2)} = \lim_{x \to -7} \frac{3-x}{x+2} = \frac{3-(-7)}{-7+2} = \frac{10}{-5} = -2$$

196. Substituting $x = 6$ directly into the expression yields $\frac{8}{0}$, which suggests the presence of a vertical asymptote for the function $f(x) = \frac{x+2}{6-x}$ at $x = 6$. Note that substituting values close to 6 from the right side into the expression $6 - x$ produces y-values that become unboundedly large in the negative direction. Hence, we conclude that $\lim_{x \to 6^+} \frac{x+2}{6-x} = -\infty$.

197. To compute this limit, first simplify the expression $(x+h)^3$, as follows:

$$(x+h)^3 = (x+h)^2(x+h) = \left(x^2 + 2hx + h^2\right)(x+h)$$

$$= x^3 + 2hx^2 + h^2x + x^2h + 2h^2x + h^3 = x^3 + 3hx^2 + 3h^2x + h^3$$

Hence, the original problem is equivalent to

$$\lim_{h \to 0} \frac{2\left(x^3 + 3hx^2 + 3h^2x + h^3\right) - 2x^3}{h}$$

Now, simplify the numerator, cancel factors that are common to both the numerator and the denominator, and then substitute $h = 0$ into the simplified expression, as follows:

$$\lim_{h \to 0} \frac{2(x+h)^3 - 2x^3}{h} = \lim_{h \to 0} \frac{2\left(x^3 + 3hx^2 + 3h^2x + h^3\right) - 2x^3}{h}$$

$$= \lim_{h \to 0} \frac{\cancel{2x^3} + 6hx^2 + 6h^2x + 2h^3 - \cancel{2x^3}}{h} = \lim_{h \to 0} \frac{h\left(6x^2 + 6hx + 2h^2\right)}{h}$$

$$= \lim_{h \to 0}\left(6x^2 + 6hx + 2h^2\right) = 6x^2$$

198. a. Substituting $x = \pi$ directly into the expression $\frac{\cos x + 1}{\cos x - 1}$ yields

$$\frac{\cos \pi + 1}{\cos \pi - 1} = \frac{-1+1}{-1-1} = \frac{0}{-2} = 0$$

Thus, $\lim_{x \to \pi} \frac{\cos x + 1}{\cos x - 1} = 0$.

199. c. To compute this limit, factor the numerator and the denominator, cancel factors that are common to both the numerator and the denominator, and substitute $x = 3$ into the simplified expression, as follows:

$$\lim_{x \to 3} \frac{x^3 + 2x^2 - 15x}{x^2 - 9} = \lim_{x \to 3} \frac{x(x-3)(x+5)}{(x-3)(x+3)} = \lim_{x \to 3} \frac{x(x+5)}{x+3} = \frac{3(3+5)}{3+3} = \frac{24}{6} = 4$$

200. a. To compute this limit, multiply the numerator and the denominator by the expression $\left(1+2\sqrt{x}\right)$. Upon simplifying the numerator, cancel factors that are common to both the numerator and the denominator, and substitute $x = \frac{1}{4}$ into the simplified expression, as follows:

$$\lim_{x \to \frac{1}{4}} \frac{1 - 2\sqrt{x}}{1 - 4x} = \lim_{x \to \frac{1}{4}} \frac{1 - 2\sqrt{x}}{1 - 4x} \cdot \frac{1 + 2\sqrt{x}}{1 + 2\sqrt{x}} = \lim_{x \to \frac{1}{4}} \frac{1 - 4x}{\left(1 - 4x\right)\left(1 + 2\sqrt{x}\right)}$$

$$= \lim_{x \to \frac{1}{4}} \frac{1}{1 + 2\sqrt{x}} = \frac{1}{1 + 2\sqrt{\frac{1}{4}}} = \frac{1}{1 + 2\left(\frac{1}{2}\right)} = \frac{1}{1 + 1} = \frac{1}{2}$$

201. b. To compute this limit, we make use of the fact that $\lim_{x \to 0} \left(\frac{\sin x}{x}\right) = 1$,

together with the rule $\lim_{x \to c} \left[f(x)g(x) \right] = \left(\lim_{x \to c} f(x) \right) \cdot \left(\lim_{x \to c} g(x) \right)$, as follows:

$$\lim_{x \to 0^{-}} \frac{\sin x \cos x}{3x} = \lim_{x \to 0^{-}} \left(\frac{\sin x}{x} \cdot \frac{\cos x}{3} \right) = \lim_{x \to 0^{-}} \left(\frac{\sin x}{x} \right) \cdot \lim_{x \to 0^{-}} \left(\frac{\cos x}{3} \right) = 1 \cdot \frac{\cos 0}{3} = \frac{1}{3}$$

202. d. To compute this limit, use the fact that $e^{2x} = \left(e^x\right)^2$ and factor the denominator as a difference of squares. Then, cancel factors that are common to both the numerator and the denominator, and substitute $x = 0$ into the simplified expression, as follows:

$$\lim_{x \to 0^{+}} \frac{1 - e^x}{1 - e^{2x}} = \lim_{x \to 0^{+}} \frac{1 - e^x}{1 - \left(e^x\right)^2} = \lim_{x \to 0^{+}} \frac{1 - e^x}{\left(1 - e^x\right)\left(1 + e^x\right)} = \lim_{x \to 0^{+}} \frac{1}{1 + e^x} = \frac{1}{1 + e^0}$$

$$= \frac{1}{1 + 1} = \frac{1}{2}$$

203. a. Substituting $x = -\frac{3\pi}{4}$ into the expression $\tan\left(x + \frac{\pi}{4}\right)$ yields $\tan\left(-\frac{\pi}{2}\right)$, which is known to be undefined. Using the graph of $y = \tan\left(x + \frac{\pi}{4}\right)$, shown here, as a guide, we observe that substituting values close to $-\frac{3\pi}{4}$ from the right side into this expression produces y-values that become unboundedly large in the negative direction. So, we conclude that $\lim\limits_{x \to \left(-\frac{3\pi}{4}\right)^+} \tan\left(x + \frac{\pi}{4}\right) = -\infty$.

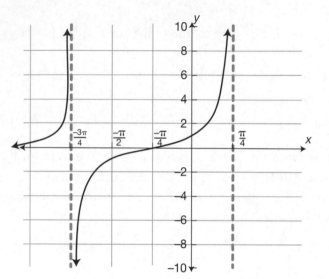

204. **c.** Substituting $x = 2e$ into the expression $\ln(x - 2e)$ yields $\ln 0$, which is known to be undefined. Using the graph of $y = \ln(x - 2e)$, shown here, as a guide, we observe that substituting values close to $2e$ from the right side into this expression produces y-values that become unboundedly large (in the negative direction). So, we conclude that
$$\lim_{x \to (2e)^+} \ln(x - 2e) = -\infty.$$

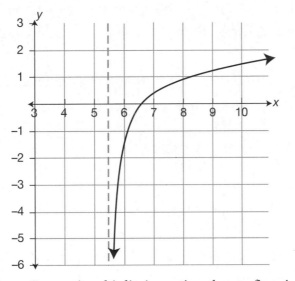

205. **a.** Computing this limit requires that we first simplify the complex fraction, as follows:

$$\frac{4-x}{\frac{1}{x-5}+1} = \frac{4-x}{\frac{1}{x-5}+\frac{x-5}{x-5}} = \frac{4-x}{\frac{1+x-5}{x-5}} = \frac{4-x}{\frac{x-4}{x-5}} = (4-x)\cdot\frac{x-5}{x-4} = -(x-4)\cdot\frac{x-5}{x-4}$$
$$= -(x-5)$$

Hence, we conclude

$$\lim_{x \to 4} \frac{4-x}{\frac{1}{x-5}+1} = \lim_{x \to 4} -(x-5) = -(4-5) = -(-1) = 1$$

206. Using the graphs, we see that $\lim_{x \to -4^+} g(x) = 0$ and $\lim_{x \to -4^+} f(x) = 5$. Now, using Arithmetic of Limits rules 3 and 4, provided at the beginning of the chapter, this yields

$$\lim_{x \to -4^+} \left(2g(x) - 3f(x)\right) = 2\lim_{x \to -4^+} g(x) - 3\lim_{x \to -4^+} f(x) = 2(0) - 3(5)$$
$$= 0 - 15 = -15$$

207. Using the graphs, we see that $\lim\limits_{x\to 0} f(x)=5$ and $\lim\limits_{x\to 0} g(x)=-2$. Now, using Arithmetic of Limits rule 5, provided at the beginning of the chapter, this yields

$$\lim\limits_{x\to 0}\big(f(x)\cdot g(x)\big)=\Big(\lim\limits_{x\to 0} f(x)\Big)\cdot\Big(\lim\limits_{x\to 0} g(x)\Big)=(5)(-2)=-10$$

208. Using the graphs, we see that $\lim\limits_{x\to 0} f(x)=5$ and $\lim\limits_{x\to 0} g(x)=-2$. Since the denominator of the expression $\frac{g(x)}{f(x)}$ becomes increasingly large while the numerator settles down to a fixed real number, we conclude that $\lim\limits_{x\to 5}\frac{g(x)}{f(x)}=0$.

209. Using the graph, we see that $\lim\limits_{x\to -3} g(x)=3$. Now, using Arithmetic of Limits rules 3 and 7, provided at the beginning of the chapter, this yields

$$\lim\limits_{x\to -3}-2\sqrt{g(x)}=-2\lim\limits_{x\to -3}\sqrt{g(x)}=-2\sqrt{\lim\limits_{x\to -3} g(x)}=-2\sqrt{3}$$

210. Using the graph, we see that $\lim\limits_{x\to 2} g(x)=0$ and for values of x close to 2 (but not at 2) on both sides, the values of $g(x)$ are negative. Hence, we conclude that $\lim\limits_{x\to 2}\frac{1}{g(x)}=-\infty$.

211. Using the graph, we see that $\lim\limits_{x\to -6^-} f(x)=0$ and for values of x close to −6 (but not at −6) on the left side, the values of $f(x)$ are positive. Hence, we conclude that $\lim\limits_{x\to -6^-}\frac{1}{f(x)}=\infty$.

212. Using the graph, we see that $\lim\limits_{x\to 3} g(x)=-1$. Now, using Arithmetic of Limits rules 3 and 7, provided at the beginning of the chapter, this yields

$$\lim\limits_{x\to 3}\sqrt[3]{8g(x)}=\sqrt[3]{\lim\limits_{x\to 3} 8g(x)}=\sqrt[3]{8\lim\limits_{x\to 3} g(x)}=\sqrt[3]{8(-1)}=-2$$

213. Using the graphs, we see that $\lim\limits_{x\to\infty} f(x)=0$ and $\lim\limits_{x\to\infty} g(x)=-1$. Now, using Arithmetic of Limits rules 3 and 4, provided at the beginning of the chapter, this yields

$$\lim\limits_{x\to\infty}\big(g(x)+2f(x)\big)=\lim\limits_{x\to\infty} g(x)+2\lim\limits_{x\to\infty} f(x)=-1+2(0)=-1$$

214. True. Using the graphs, we see that $\lim\limits_{x \to 2} f(x) = 2$ and $\lim\limits_{x \to 2} g(x) = 0$.

Hence, using the Arithmetic of Limits rules provided at the beginning of the chapter yields

$$\lim_{x \to 2}\left[f(x)\right]^3 = \left[\lim_{x \to 2} f(x)\right]^3 = 2^3 = 8$$

$$\lim_{x \to 2}\left(2g(x)+1\right) = 2\lim_{x \to 2} g(x) + \lim_{x \to 2} 1 = 2(0) + 1 = 1$$

Applying Arithmetic of Limits rule 6 then yields

$$\lim_{x \to 2} h(x) = \lim_{x \to 2} \frac{[f(x)]^3}{2g(x)+1} = \frac{\lim\limits_{x \to 2}[f(x)]^3}{\lim\limits_{x \to 2}(2g(x)+1)} = \frac{8}{1} = 8$$

Since this value coincides with $h(2)$, we conclude that h is continuous at $x = 2$.

215. False. Using the graph, we see that $\lim\limits_{x \to -4^-} f(x) = -3$ and $\lim\limits_{x \to -4^+} f(x) = 5$.

Since these values are different, we conclude that f does not have a limit at $x = -4$.

Differentiation Problems

Measuring the instantaneous rate of change of one quantity with respect to another is the primary focus of differential calculus. The notion of an *average rate of change* is familiar since such a quantity is used to compute the slope of a line. Computing the rate at which a function $y = f(x)$ changes with respect to x at a specific value $x = a$ requires that we compute a limit of certain average rates involving functional values close to a. This notion, known as the *derivative* of f at $x = a$, is defined next, and several computational rules are provided.

    Definition: The *derivative* of $y = f(x)$ at $x = a$ is $\lim_{h \to 0} \frac{f(a+h) - f(a)}{h}$, denoted by $f'(a)$.

The following steps are used to compute a derivative using this definition:

1. Form $f(a+h)$; that is, substitute $a+h$ for x in the formula for $f(x)$.
2. Subtract $f(a)$ to obtain $f(a+h)-f(a)$.
3. Divide by h to obtain $\frac{f(a+h)-f(a)}{h}$, which is called the *difference quotient*. Note this quantity is the slope of the chord line connecting $(a, f(a))$ to $(a+h, f(a+h))$.
4. Take the limit of the difference quotient as $h \to 0$; that is, compute $\lim\limits_{h \to 0} \frac{f(a+h)-f(a)}{h}$.

If this limit exists, it is called the *derivative* of $f(x)$ at a and is denoted by $f'(a)$. We say in such case that $f(x)$ is *differentiable* at a (which merely means the derivative exists at a). If the limit does not exist, we say $f(x)$ is not differentiable at a and there is no derivative there. Alternative notation used to denote the derivative of $y = f(x)$ that means the same thing as $f'(x)$ is the so-called *Leibniz notation* $\frac{dy}{dx}$. Geometrically, $f'(a)$ is the slope of the tangent line to the graph of $f(x)$ at a.

The process of computing the derivative of a function $f(x)$ produces another function $f'(x)$, defined formally by

$$f'(x) = \lim_{h \to 0} \frac{f(x+h)-f(x)}{h}$$

for any x for which this limit exists and is finite.

This process can be applied again to $f'(x)$, resulting in the *second derivative* of f, denoted by $f''(x)$ or $\frac{d^2 y}{dx^2}$. Likewise, the process can be applied again and again to form the third, fourth, and so on derivatives of f.

Differentiation Rules

1. **Power rule:** $(x^r)' = rx^{r-1}$
2. **Factor out constants:** $[cf(x)]' = cf'(x)$
3. **Sum/difference rule:** $[f(x) \pm g(x)]' = f'(x) \pm g'(x)$
4. **Product rule:** $[f(x)g(x)]' = f'(x)g(x) + f(x)g'(x)$
   For more factors, this extends to $(fgh)' = f'gh + fg'h + fgh'$, and so on.
5. **Quotient rule:** $\left[\dfrac{f(x)}{g(x)}\right]' = \dfrac{g(x)f'(x) - f(x)g'(x)}{[g(x)]^2}$

1

6. **Chain rule:** If $\underbrace{y = y(u)}_{y \text{ is some combination of } u}$ where $\underbrace{u = u(x)}_{u, \text{ in turn, is some combination of } x}$, then

$$\underbrace{\frac{dy}{dx}}_{\substack{\text{rate of}\\\text{change of}\\y \text{ wrt } x}} = \underbrace{\frac{dy}{du}}_{\substack{\text{rate of}\\\text{change of}\\y \text{ wrt } u}} \cdot \underbrace{\frac{du}{dx}}_{\substack{\text{rate of}\\\text{change of}\\u \text{ wrt } x}} \text{ (wrt stands for } \textit{with regard to}\text{).}$$

Some Common Derivatives

1. $\frac{d}{dx}\sin x = \cos x$

2. $\frac{d}{dx}\cos x = -\sin x$

3. $\frac{d}{dx}\tan x = \sec^2 x$

4. $\frac{d}{dx}\cot x = -\csc^2 x$

5. $\frac{d}{dx}\sec x = \sec x \tan x$

6. $\frac{d}{dx}\csc x = -\csc x \cot x$

7. $\frac{d}{dx}e^x = e^x$

8. $\frac{d}{dx}\ln x = \frac{1}{x}$

9. $\frac{d}{dx}\arcsin x = \frac{1}{\sqrt{1-x^2}}$

10. $\frac{d}{dx}\arctan x = \frac{1}{1+x^2}$

Questions

For Questions 216 through 230, compute $f'(x)$.

216. $f(x) = \frac{-5}{x^9}$

217. $f(x) = -x^5 - 4x^4 + 5x^3 - 2x^2 + 9x - 1$

218. $f(x) = 5 + \frac{2}{x} - \frac{1}{x^2}$

219. $f(x) = -4\sqrt{x} + 8\sqrt[3]{x}$

220. $f(x) = 4x^5 e^x$

221. $f(x) = \sin x \cos x$

222. $f(x) = \frac{x \ln x}{e^x}$

223. $f(x) = \left(x^7 + 8x^5 - 2x - 1\right)^{-9}$

224. $f(x) = \sqrt[5]{-3e^x + 2}$

225. $f(x) = \cos\left(x^3\right)$

226. $f(x) = \cos^3 x$

227. $f(x) = \frac{\tan(2x)}{\cos(3x)}$

228. $f(x) = \sec\left(\ln(2x) + x\sqrt{x}\right)$

229. $f(x) = \cos^{\frac{5}{3}}\left(\sqrt{\ln\left(x^2 - 3x\right)}\right)$

230. $f(x) = \sin(\sin(\sin(\sin(\sin(x)))))$

231. Compute the second derivative of $f(x) = \frac{x-3}{x+3}$.

232. Determine the equation of the tangent line to the graph of $f(x) = x\sin x$ at $x = \frac{\pi}{4}$.

For Questions 233 through 235, use implicit differentiation to compute $\frac{dy}{dx}$.

233. $\left(y + 1\right)^5 = 3x^5 - 4x$

234. $y = \sqrt{2x - 3y^2}$

235. $-2xy^3 - e^{2x} = \ln\left(xy\right)$

For Questions 236 through 242, assume that the domain of the function f is the set of all real numbers and refer to the following graph of its derivative $y = f'(x)$.

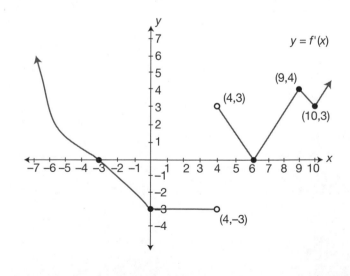

236. True or false? $f(x)$ is not differentiable at $x = 4$.

237. True or false? The graph of $y = f(x)$ is decreasing on $(-\infty, 0) \cup (4, 6)$.

238. At which of the following x-values must $f(x)$ have a local maximum?
   a. −3
   b. −3 and 6
   c. −3, 6, and 9
   d. −3, 6, 9, and 10

239. On which of the following sets is the graph of $y = f(x)$ concave down?
   a. $(6, 9) \cup (10, \infty)$
   b. $(-\infty, 0) \cup (4, 6) \cup (9, 10)$
   c. $(-3, 4)$
   d. $(-\infty, -3) \cup (4, \infty)$

240. True or false? The graph of $y = f(x)$ is constant on $(0, 4)$.

241. At which of the following x-values is the graph of $y = f'(x)$ discontinuous?
   a. 6, 9, and 10 only
   b. −3 and 6 only
   c. 4 only
   d. −3, 4, 6, 9, and 10 only

242. True or false? The second derivative of $y = f(x)$ does not exist when $x = 4, 6, 9,$ or 10.

243. Which of the following is the equation of the tangent line to the implicitly defined function $y = \sin(xy)$ at the point $\left(\frac{\pi}{4}, 0\right)$?
   a. There is no such tangent line.
   b. $y = x - \frac{\pi}{4}$
   c. $x = \frac{\pi}{4}$
   d. $y = 0$

244. On which of the following sets is the graph of $f(x) = \arctan(2x)$ concave up?

a. $\mathbb{R}$

b. $\left(-\frac{\pi}{8}, \frac{\pi}{8}\right)$

c. $(0, \infty)$

d. $(-\infty, 0)$

245. On which of the following intervals is the graph of $f(x) = xe^{-2x}$ increasing?

a. $(-\infty, -1)$

b. $\left(-\infty, \frac{1}{2}\right)$

c. $(-1, \infty)$

d. $\left(\frac{1}{2}, \infty\right)$

Answers

216. Since $f(x) = -5x^{-9}$, we see that $f'(x) = -5(-9)x^{-10} = \frac{45}{x^{10}}$.

217. $f'(x) = -5x^4 - 4(4)x^3 + 5(3)x^2 - 2(2)x^1 + 9 = -5x^4 - 16x^3 + 15x^2 - 4x + 9$

218. Since $f(x) = 5 + 2x^{-1} - x^{-2}$, we see that $f'(x) = 2(-1)x^{-2} - (-2)x^{-3}$
$= -\frac{2}{x^2} + \frac{2}{x^3}$.

219. Since $f(x) = -4x^{\frac{1}{2}} + 8x^{\frac{1}{3}}$, we see that $f'(x) = -4\left(\frac{1}{2}\right)x^{-\frac{1}{2}} + 8\left(\frac{1}{3}\right)x^{-\frac{2}{3}}$
$= -\frac{2}{\sqrt{x}} + \frac{2}{\sqrt[3]{x^2}}$.

220. Applying the product rule yields $f'(x) = \left(4x^5\right)e^x + \left(20x^4\right)e^x$
$= 4x^4(x+5)e^x$.

221. Applying the product rule and a known trigonometric identity when simplifying the last step yields
$f'(x) = \sin x(-\sin x) + (\cos x)\cos x = \cos^2 x - \sin^2 x = \cos 2x$

222. Applying the quotient rule first, and then the product rule when differentiating the expression $x \ln x$, yields

$f'(x) = \dfrac{\left(e^x\right)(x \ln x)' - (x \ln x)\left(e^x\right)'}{\left(e^x\right)^2}$

$= \dfrac{e^x\left(x \cdot \frac{1}{x} + \ln x \cdot 1\right) - (x \ln x)\left(e^x\right)}{\left(e^x\right)^2}$

$= \dfrac{e^x\left(1 + \ln x - x \ln x\right)}{\left(e^x\right)^2}$

$= \dfrac{1 + \ln x - x \ln x}{e^x}$

223. Applying the chain rule yields
$f'(x) = -9\left(x^7 + 8x^5 - 2x - 1\right)^{-10}\left(x^7 + 8x^5 - 2x - 1\right)'$

$= -9\left(x^7 + 8x^5 - 2x - 1\right)^{-10}\left(7x^6 + 40x^4 - 2\right)$

$= \dfrac{-9\left(7x^6 + 40x^4 - 2\right)}{\left(x^7 + 8x^5 - 2x - 1\right)^{10}}$

224. Since $f(x) = \left(-3e^x + 2\right)^{\frac{1}{5}}$, applying the chain rule yields

$$f'(x) = \left(\tfrac{1}{5}\right)\left(-3e^x + 2\right)^{-\frac{4}{5}}\left(-3e^x + 2\right)'$$

$$= \left(\tfrac{1}{5}\right)\left(-3e^x + 2\right)^{-\frac{4}{5}}\left(-3e^x\right)$$

$$= \frac{-3e^x}{5\left(-3e^x + 2\right)^{\frac{4}{5}}}$$

225. Applying the chain rule yields

$$f'(x) = -\sin\left(x^3\right)\cdot\left(x^3\right)' = -3x^2\sin\left(x^3\right)$$

226. Since $\cos^3 x = \left(\cos x\right)^3$, applying the chain rule yields

$$f'(x) = 3\left(\cos x\right)^2\left(\cos x\right)' = 3\left(\cos x\right)^2\left(-\sin x\right) = -3\sin x \cos^2 x$$

227. Applying the quotient rule first, and then the chain rule when differentiating the individual expression comprising the numerator and the denominator, yields

$$f'(x) = \frac{\cos(3x)\left(\tan(2x)\right)' - \tan(2x)\left(\cos(3x)\right)'}{\cos^2(3x)}$$

$$= \frac{\cos(3x)\left(2\sec^2(2x)\right) - \tan(2x)\left(-3\sin(3x)\right)}{\cos^2(3x)}$$

$$= \frac{2\cos(3x)\sec^2(2x) + 3\sin(3x)\tan(2x)}{\cos^2(3x)}$$

228. Successive applications of the chain rule yield

$$f'(x) = \left[\sec\left(\ln(2x) + x\sqrt{x}\right)\tan\left(\ln(2x) + x\sqrt{x}\right)\right]\cdot\left(\ln(2x) + \underbrace{x\sqrt{x}}_{=x^{\frac{3}{2}}}\right)'$$

$$= \left[\sec\left(\ln(2x) + x\sqrt{x}\right)\tan\left(\ln(2x) + x\sqrt{x}\right)\right]\cdot\left(\tfrac{1}{2x}(2) + \tfrac{3}{2}x^{\frac{1}{2}}\right)$$

$$= \left[\sec\left(\ln(2x) + x\sqrt{x}\right)\tan\left(\ln(2x) + x\sqrt{x}\right)\right]\cdot\left(\tfrac{1}{x} + \tfrac{3}{2}\sqrt{x}\right)$$

229. Since $f(x) = \left(\cos\left(\sqrt{\ln(x^2 - 3x)}\right)\right)^{\frac{5}{3}}$, successive applications of the chain rule yield

$$f'(x) = \tfrac{5}{3}\left(\cos\left(\sqrt{\ln(x^2 - 3x)}\right)\right)^{\frac{2}{3}}\left(\cos\left(\sqrt{\ln(x^2 - 3x)}\right)\right)'$$

$$= \tfrac{5}{3}\left(\cos\left(\sqrt{\ln(x^2 - 3x)}\right)\right)^{\frac{2}{3}}\left(-\sin\left(\sqrt{\ln(x^2 - 3x)}\right)\right)\left(\sqrt{\ln(x^2 - 3x)}\right)'$$

$$= \tfrac{5}{3}\left(\cos\left(\sqrt{\ln(x^2 - 3x)}\right)\right)^{\frac{2}{3}}\left(-\sin\left(\sqrt{\ln(x^2 - 3x)}\right)\right)$$

$$\left(\tfrac{1}{2}(\ln(x^2 - 3x))^{-\frac{1}{2}}\right)\left(\ln(x^2 - 3x)\right)'$$

$$= \tfrac{5}{3}\left(\cos\left(\sqrt{\ln(x^2 - 3x)}\right)\right)^{\frac{2}{3}}\left(-\sin\left(\sqrt{\ln(x^2 - 3x)}\right)\right)$$

$$\left(\tfrac{1}{2}\ln(x^2 - 3x))^{-\frac{1}{2}}\right)\left(\tfrac{1}{x^2 - 3x}\right)\left(x^2 - 3x\right)'$$

$$= \tfrac{5}{3}\left(\cos\left(\sqrt{\ln(x^2 - 3x)}\right)\right)^{\frac{2}{3}}\left(-\sin\left(\sqrt{\ln(x^2 - 3x)}\right)\right)$$

$$\left(\tfrac{1}{2}(\ln(x^2 - 3x))^{-\frac{1}{2}}\right)\left(\tfrac{1}{x^2 - 3x}\right)(2x - 3)$$

230. Successive applications of the chain rule yield

$$f'(x) = \cos(\sin(\sin(\sin(\sin(x))))) \cdot \left[\sin(\sin(\sin(\sin(x))))\right]'$$

$$= \cos(\sin(\sin(\sin(\sin(x))))) \cdot \cos(\sin(\sin(\sin(x))))$$

$$\cdot \left[\sin(\sin(x))\right]'$$

$$= \cos(\sin(\sin(\sin(\sin(x))))) \cdot \cos(\sin(\sin(\sin(x))))$$

$$\cdot \cos(\sin(\sin(x))) \cdot \left[\sin(\sin(x))\right]'$$

$$= \cos(\sin(\sin(\sin(\sin(x))))) \cdot \cos(\sin(\sin(\sin(x))))$$

$$\cdot \cos(\sin(\sin(x))) \cdot \cos(\sin(x)) \cdot \left[\sin(x)\right]'$$

$$= \cos(\sin(\sin(\sin(\sin(x))))) \cdot \cos(\sin(\sin(\sin(x))))$$

$$\cdot \cos(\sin(\sin(x))) \cdot \cos(\sin(x)) \cdot \cos(x)$$

231. Applying the quotient rule yields

$$f'(x) = \frac{(x+3)(1)-(x-3)(1)}{(x+3)^2} = \frac{x+3-x+3}{(x+3)^2} = \frac{6}{(x+3)^2} = 6(x+3)^{-2}$$

Now, applying the chain rule to differentiate $f'(x)$ yields

$$f''(x) = 6(-2)(x+3)^{-3}(1) = -\frac{12}{(x+3)^3}$$

232. We need the slope of this line and the point of tangency, which must lie on the tangent line. The slope is $f'\left(\frac{\pi}{4}\right)$ and the point of tangency is $\left(\frac{\pi}{4}, f\left(\frac{\pi}{4}\right)\right)$.

Observe that

$$f'(x) = x(\cos x) + (1)\sin x = x\cos x + \sin x$$

$$f'\left(\frac{\pi}{4}\right) = \frac{\pi}{4}\cos\left(\frac{\pi}{4}\right) + \sin\left(\frac{\pi}{4}\right) = \frac{\pi}{4}\left(\frac{\sqrt{2}}{2}\right) + \frac{\sqrt{2}}{2} = \frac{\sqrt{2}}{2}\left(\frac{\pi}{4} + 1\right) = \frac{\sqrt{2}(\pi + 4)}{8}$$

Also,

$$f\left(\frac{\pi}{4}\right) = \left(\frac{\pi}{4}\right)\sin\left(\frac{\pi}{4}\right) = \left(\frac{\pi}{4}\right)\left(\frac{\sqrt{2}}{2}\right) = \frac{\pi\sqrt{2}}{8}$$

So, the point of tangency is $\left(\frac{\pi}{4}, \frac{\pi\sqrt{2}}{8}\right)$. Hence, using the point-slope formula for a line, we conclude that the equation of the desired tangent line is

$$y - \frac{\pi\sqrt{2}}{8} = \frac{\sqrt{2}(\pi + 4)}{8}\left(x - \frac{\pi}{4}\right)$$

233. Differentiation of both sides with respect to x, applying the chain rule as necessary, yields

$$5(y+1)^4 \frac{dy}{dx} = 15x^4 - 4$$

$$\frac{dy}{dx} = \frac{15x^4 - 4}{5(y+1)^4}$$

234. Differentiation of both sides with respect to x, applying the chain rule as necessary, yields

$$\frac{dy}{dx} = \frac{1}{2}\left(2x - 3y^2\right)^{-\frac{1}{2}}\left(2 - 6y\frac{dy}{dx}\right)$$

$$\frac{dy}{dx} = \left(2x - 3y^2\right)^{-\frac{1}{2}} - 3y\left(2x - 3y^2\right)^{-\frac{1}{2}}\frac{dy}{dx}$$

$$\frac{dy}{dx} + 3y\left(2x - 3y^2\right)^{-\frac{1}{2}}\frac{dy}{dx} = \left(2x - 3y^2\right)^{-\frac{1}{2}}$$

$$\frac{dy}{dx}\left[1 + 3y\left(2x - 3y^2\right)^{-\frac{1}{2}}\right] = \left(2x - 3y^2\right)^{-\frac{1}{2}}$$

$$\frac{dy}{dx} = \frac{\left(2x - 3y^2\right)^{-\frac{1}{2}}}{1 + 3y\left(2x - 3y^2\right)^{-\frac{1}{2}}}$$

We can simplify the right side even further as follows:

$$\frac{dy}{dx} = \frac{\dfrac{1}{\left(2x - 3y^2\right)^{\frac{1}{2}}}}{1 + \dfrac{3y}{\left(2x - 3y^2\right)^{\frac{1}{2}}}} = \frac{\dfrac{1}{\left(2x - 3y^2\right)^{\frac{1}{2}}}}{\dfrac{\left(2x - 3y^2\right)^{\frac{1}{2}} + 3y}{\left(2x - 3y^2\right)^{\frac{1}{2}}}} = \frac{1}{\left(2x - 3y^2\right)^{\frac{1}{2}}} \cdot \frac{\left(2x - 3y^2\right)^{\frac{1}{2}}}{\left(2x - 3y^2\right)^{\frac{1}{2}} + 3y}$$

$$= \frac{1}{\left(2x - 3y^2\right)^{\frac{1}{2}} + 3y}$$

235. Differentiation of both sides with respect to x, applying the chain rule as necessary, yields

$$-2x\left(3y^2\frac{dy}{dx}\right)+y^3(-2)-2e^{2x}=\frac{1}{xy}\left(x\frac{dy}{dx}+y(1)\right)$$

$$-6xy^2\frac{dy}{dx}-2y^3-2e^{2x}=\frac{1}{y}\frac{dy}{dx}+\frac{1}{x}$$

$$-6xy^2\frac{dy}{dx}-\frac{1}{y}\frac{dy}{dx}=\frac{1}{x}+2y^3+2e^{2x}$$

$$-\frac{dy}{dx}\left(6xy^2+\frac{1}{y}\right)=\frac{1}{x}+2y^3+2e^{2x}$$

$$\frac{dy}{dx}=-\frac{\frac{1}{x}+2y^3+2e^{2x}}{6xy^2+\frac{1}{y}}$$

We can simplify this expression further, as follows:

$$\frac{dy}{dx}=-\frac{\frac{1}{x}+y^3+2e^{2x}}{6xy^2+\frac{1}{y}}\cdot\frac{xy}{xy}=\frac{y+xy^3+2xe^{2x}}{6x^2y^3+x}$$

236. True. The graph of $y=f'(x)$ has only open holes at $x=4$, so $f'(4)$ is not defined. Thus, f is not differentiable at $x=4$.

237. False. The graph of $y=f(x)$ is decreasing at precisely those x-values at which the graph of $y=f'(x)$ is below the x-axis, namely on the interval $(-3,4)$.

238. a. The graph of $y=f(x)$ has a local maximum at $x=a$ if $y=f'(x)>0$ for x-values very close to a on the left and $y=f'(x)<0$ for x-values very close to a on the right. This situation occurs only at $x=-3$.

239. b. The graph of $y=f(x)$ is concave down at those x-values at which the graph of $y=f'(x)$ is decreasing (i.e., on intervals where the y-values get smaller from left to right), namely $(-\infty,0)\cup(4,6)\cup(9,10)$.

240. False. The graph of $y=f(x)$ is constant on an interval I if and only if $f'(x)=0$ at every x-value in I. From the given graph, we see that $f'(x)=-3<0$ on (0,4).

241. c. The graph of $y=f'(x)$ is discontinuous only at $x=4$, due to the jump in the graph.

242. True. The second derivative of $y=f(x)$ does not exist at those x-values at which the graph of $y=f'(x)$ is discontinuous or has a sharp corner. This happens when $x=4, 6, 9,$ and 10.

243. d. We need the slope of this line and the point of tangency, which must lie on the tangent line. The slope is the value of $\frac{dy}{dx}$ at the point of tangency $\left(\frac{\pi}{4}, 0\right)$. Using implicit differentiation yields

$$\frac{dy}{dx} = \cos(xy) \cdot \left(x\frac{dy}{dx} + y(1)\right)$$

$$\frac{dy}{dx} = x\cos(xy)\frac{dy}{dx} + y\cos(xy)$$

$$\frac{dy}{dx} - x\cos(xy)\frac{dy}{dx} = y\cos(xy)$$

$$\frac{dy}{dx}\big(1 - x\cos(xy)\big) = y\cos(xy)$$

$$\frac{dy}{dx} = \frac{y\cos(xy)}{1 - x\cos(xy)}$$

Evaluating this expression at the point $\left(\frac{\pi}{4}, 0\right)$ yields

$$\frac{dy}{dx} = \frac{0 \cdot \cos(0)}{1 - \left(\frac{\pi}{4}\right)\cos(0)} = 0$$

Therefore, the tangent line is the horizontal line through the point $\left(\frac{\pi}{4}, 0\right)$, namely $y = 0$.

244. d. The graph of $f(x) = \arctan(2x)$ is concave up on those intervals where $f''(x) > 0$. We compute the first and second derivatives of f, as follows:

$$f'(x) = \frac{1}{1 + (2x)^2} \cdot (2x)' = \frac{2}{1 + 4x^2} = 2\big(1 + 4x^2\big)^{-1}$$

$$f''(x) = -2\big(1 + 4x^2\big)^{-2}\big(1 + 4x^2\big)' = -2\big(1 + 4x^2\big)^{-2}(8x) = \frac{-16x}{\big(1 + 4x^2\big)^2}$$

Since the denominator of $f''(x)$ is always nonnegative, the only x-values for which $f''(x) > 0$ are those for which the numerator is positive. This happens for only those x-values in the interval $(-\infty, 0)$.

245. b. The graph of $f(x) = xe^{-2x}$ is increasing on those intervals where $f'(x) > 0$. We compute the derivative of f, as follows:

$$f'(x) = x \cdot \big(-2e^{-2x}\big) + (1)\big(e^{-2x}\big) = e^{-2x}(-2x + 1)$$

Since e^{-2x} is always positive, the only x-values for which $f'(x) > 0$ are those for which $-2x + 1 > 0$; that is, for only those x-values in the interval $\left(-\infty, \frac{1}{2}\right)$.

10

Applications of Differentiation Problems I: Related Rates

The technique of implicit differentiation is particularly useful when studying applications in which the formula relating the quantities of interest (e.g., the formula for the volume of a sphere with radius a, $V = \frac{4}{3}\pi a^3$; the formula for the area of a circle with radius r, $A = \pi r^2$; and the Pythagorean theorem relating the legs a and b to the hypotenuse c of a right triangle, $a^2 + b^2 = c^2$). Specifically, if one of the quantities in the formula changes with a variable, like time, then all other quantities in the formula change accordingly. Hence, differentiating both sides of the formula with respect to this variable yields an equation that relates the rates of the quantities involved. We investigate such applications in this chapter.

Questions

For Questions 246 through 250, assume that all quantities involved depend on the variable t (think of t as time). Differentiate each of the following with respect to t.

246. $y = \left(x^3 + x - 1\right)^5$

247. $y^4 - 3x^2 = \cos(y)$

248. $z = \frac{2}{5}x^2 + \frac{2}{5}y^2 + \frac{3}{5x}$

249. $V = \frac{4}{3}\pi r^3$

250. $A = \frac{1}{2}bh$

251. Suppose $xy^2 = x^2 + 3$. What is $\frac{dy}{dt}$ when $\frac{dx}{dt} = 8$, $x = 3$, and $y = -2$?

252. Suppose $A^3 = B^2 + 4C^2$, $\frac{dA}{dt} = 8$, and $\frac{dC}{dt} = -2$. What is $\frac{dB}{dt}$ when $A = 2$, $B = 2$, and $C = 1$?

253. Suppose $A = I^2 + 6R$. If I increases by 4 feet per minute and R increases by 2 square feet every minute, how fast is A changing when $I = 20$?

254. The height of a triangle increases by 2 feet every minute while its base shrinks by 6 feet every minute. How fast is the area of the triangle changing when the height is 15 feet and the base is 20 feet?

255. The surface area of a sphere with radius r is $A = 4\pi r^2$. If the radius is decreasing by 2 inches every minute, how fast is the surface area shrinking when the radius is 20 inches?

256. A circle increases in area by 20 square feet every hour. How fast is the radius increasing when the radius is 4 feet?

257. The volume of a cube grows by 1,200 cubic inches every minute. How fast is each side growing when each side is 10 inches?

258. The height of a triangle grows by 5 inches each hour. The area is increasing by 100 square inches each hour. How fast is the base of the triangle increasing when the height is 20 inches and the base is 12 inches?

259. One end of a 10-foot-long board is lifted straight off the ground at 1 foot per second. How fast will the other end drag along the ground after 6 seconds?

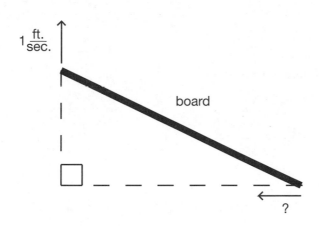

260. A kite is 100 feet off the ground and moving horizontally at 13 feet per second. How quickly must the string be let out when the string is 260 feet long?

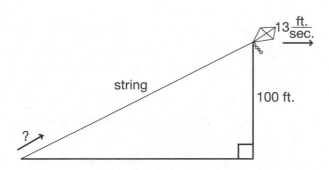

Answers

246. $\frac{dy}{dt} = 5\left(x^3 + x - 1\right)^4 \cdot \left(3x^2 \frac{dx}{dt} + \frac{dx}{dt}\right) = 5\left(x^3 + x - 1\right)^4 \cdot \left(3x^2 + 1\right)\frac{dx}{dt}$

247. $4y^3 \frac{dy}{dt} - 6x\frac{dx}{dt} = -\sin(y)\frac{dy}{dt}$

248. $\frac{dz}{dt} = \frac{4}{5}x \cdot \frac{dx}{dt} + \frac{4}{5}y \cdot \frac{dy}{dt} - \frac{3}{5x^2} \cdot \frac{dx}{dt}$

249. $\frac{dV}{dt} = 4\pi r^2 \cdot \frac{dr}{dt}$

250. $\frac{dA}{dt} = \frac{1}{2} \cdot \frac{db}{dt} \cdot h + \frac{dh}{dt} \cdot \frac{1}{2}b = \frac{1}{2}\left(\frac{db}{dt} \cdot h + \frac{dh}{dt} \cdot b\right)$

251. First, implicit differentiation yields

$$x \cdot 2y\left(\frac{dy}{dt}\right) + \left(\frac{dx}{dt}\right)y^2 = 2x\left(\frac{dx}{dt}\right)$$

$$2xy\left(\frac{dy}{dt}\right) = \left(2x - y^2\right)\left(\frac{dx}{dt}\right)$$

$$\frac{dy}{dt} = \frac{(2x - y^2)\left(\frac{dx}{dt}\right)}{2xy}$$

Now, substituting the given information into the preceding expression for $\frac{dy}{dt}$ yields

$$\frac{dy}{dt} = \frac{\left(2(3) - (-2)^2\right)(8)}{2(3)(-2)} = \frac{16}{-12} = -\frac{4}{3}$$

252. First, implicit differentiation yields

$$3A^2\left(\frac{dA}{dt}\right) = 2B\left(\frac{dB}{dt}\right) + 8C\left(\frac{dC}{dt}\right)$$

$$3A^2\left(\frac{dA}{dt}\right) - 8C\left(\frac{dC}{dt}\right) = 2B\left(\frac{dB}{dt}\right)$$

$$\frac{3A^2\left(\frac{dA}{dt}\right) - 8C\left(\frac{dC}{dt}\right)}{2B} = \frac{dB}{dt}$$

Now, substituting the given information into the preceding expression for $\frac{dB}{dt}$ yields

$$\frac{dB}{dt} = \frac{3(2)^2(8) - 8(1)(-2)}{2(2)} = \frac{112}{4} = 28$$

253. First, implicit differentiation yields

$$\frac{dA}{dt} = 2I\left(\frac{dI}{dt}\right) + 6\left(\frac{dR}{dt}\right)$$

Suppressing units, we are given $\frac{dI}{dt} = 4, \frac{dR}{dt} = 2,$ and are asked to determine $\frac{dA}{dt}$ at the instant in time when $I = 20$. Substituting these values into the expression for $\frac{dA}{dt}$ yields $\frac{dA}{dt} = 172$, meaning that A is increasing at the rate of 172 square feet per minute at this particular instant.

254. The formula for the area of a triangle with base b and height h is $A = \frac{1}{2}bh$. Implicitly differentiating both sides with respect to time t yields

$$\frac{dA}{dt} = \frac{1}{2} \cdot \frac{db}{dt} \cdot h + \frac{dh}{dt} \cdot \frac{1}{2}b$$

Suppressing units, we are given $\frac{dh}{dt} = 2, \frac{db}{dt} = -6,$ and are asked to determine $\frac{dA}{dt}$ at the instant in time when $h = 15$ and $b = 20$. Substituting these values into the expression for $\frac{dA}{dt}$ yields $\frac{dA}{dt} = -25$, meaning that the area is decreasing at the rate of 25 square feet per minute at this particular instant.

255. The formula for the surface area of a sphere with radius r is $A = 4\pi r^2$. Implicitly differentiating both sides with respect to time t yields

$$\frac{dA}{dt} = 4\pi(2r)\left(\frac{dr}{dt}\right) = 8\pi r\left(\frac{dr}{dt}\right)$$

Suppressing units, we are given $\frac{dr}{dt} = -2$ and are asked to determine $\frac{dA}{dt}$ at the instant in time when $r = 20$. Substituting these values into the expression for $\frac{dA}{dt}$ yields $\frac{dA}{dt} = -320\pi \frac{\text{in.}^2}{\text{min.}}$, meaning that the area is shrinking by 320π square inches per minute at this particular instant.

256. The formula for the area of a circle with radius r is $A = \pi r^2$. Implicitly differentiating both sides with respect to time t yields

$$\frac{dA}{dt} = \pi(2r)\left(\frac{dr}{dt}\right) = 2\pi r\left(\frac{dr}{dt}\right)$$

Suppressing units, we are given $\frac{dA}{dt} = 20$ and are asked to determine $\frac{dr}{dt}$ at the instant in time when $r = 4$. Substituting these values into the expression for $\frac{dr}{dt}$ yields $\frac{dr}{dt} = \frac{5}{2\pi}$, meaning that the radius is growing at $\frac{5}{2\pi} \approx 0.796$ feet per hour at this particular instant.

257. The formula for the volume of a cube with side s is $V = s^3$. Implicitly differentiating both sides with respect to time t yields

$$\frac{dV}{dt} = 3s^2 \left(\frac{ds}{dt} \right)$$

Suppressing units, we are given $\frac{dV}{dt} = 1,200$ and are asked to determine $\frac{ds}{dt}$ at the instant in time when $s = 10$. Substituting these values into the expression for $\frac{ds}{dt}$ yields $\frac{ds}{dt} = 4$, meaning that each side is growing at the rate of 4 inches per minute at this particular instant.

258. The formula for the area of a triangle with base b and height h is $A = \frac{1}{2}bh$. Implicitly differentiating both sides with respect to time t yields

$$\frac{dA}{dt} = \frac{1}{2} \cdot \frac{db}{dt} \cdot h + \frac{dh}{dt} \cdot \frac{1}{2} b$$

Suppressing units, we are given $\frac{dA}{dt} = 100, \frac{dh}{dt} = 5$, and are asked to determine $\frac{db}{dt}$ at the instant in time when $h = 20$ and $b = 12$. Substituting these values into the expression for $\frac{db}{dt}$ yields $\frac{db}{dt} = 7$, meaning that the base is increasing at the rate of 7 inches per hour at this particular instant.

259. We model this situation using a right triangle with height is y and base x. Using the Pythagorean theorem enables us to relate the sides of the triangle by $x^2 + y^2 = 10^2$. Implicitly differentiating both sides with respect to time t yields

$$2x\frac{dx}{dt} + 2y\frac{dy}{dt} = 0$$

Note that after 6 seconds, $y = 6$ and $x^2 + 6^2 = 100$, so that $x = 8$. We are also given that $\frac{dy}{dt} = 1$. Substituting all of this information into the preceding expression involving $\frac{dx}{dt}$ yields $\frac{dx}{dt} = -\frac{3}{4}$, meaning that the end of the board is moving at the rate of $\frac{3}{4}$ feet per second along the ground at this particular instant.

260. We model this situation using a right triangle with base x and the hypotenuse (length of the string) s. Using the Pythagorean theorem enables us to relate the sides of the triangle by $x^2 + 100^2 = s^2$. Implicitly differentiating both sides with respect to time t yields

$$2x\frac{dx}{dt} = 2s\frac{ds}{dt}$$

Note that when $s = 260$, $x = 240$. We are also given that $\frac{dx}{dt} = 13$. Substituting all of this information into the preceding expression involving $\frac{ds}{dt}$ yields $\frac{ds}{dt} = 12$, meaning that the string must be let out at 12 feet per second at this particular instant.

11

Applications of Differentiation Problems II: Optimization

Knowing the minimum and maximum points of a function is useful for graphing, but even more useful in real-life situations. Businesses want to maximize their profits, builders want to minimize their costs, and drivers want to minimize distances. If we can represent the situation with a function, then the derivative will help us to identify the optimal point.

If the derivative is zero or undefined at exactly one point, then this is very likely to be the optimal point. The first derivative test states that if the function increases—that is, $f'(x) > 0$—near the point on the left side and decreases—that is $f'(x) < 0$—near the point on the right side, then a local maximum occurs at that point. Similarly, if the function decreases near the point on the left side and increases near the point on the right side, then a local minimum occurs at that point.

If there are several points of slope zero and the domain of consideration for the function is a closed, bounded interval, then plug all the *critical points* (points of slope zero, points of undefined derivative, and the two endpoints of the interval) into the original function. The *absolute maximum* of the function on this interval occurs at the *x*-value for which the largest *y*-value occurs, while the *absolute minimum* occurs at the *x*-value for which the smallest *y*-value occurs.

Questions

For Questions 261 through 265, identify the locations of all local minima and maxima of the given function on the indicated domain using the first derivative test.

261. $f(x) = \frac{2}{3}x^3 + \frac{1}{2}x^2 - 3x + 4$, x is any real number

262. $f(x) = \frac{x^2}{x^2 + 1}$, x is any real number

263. $f(x) = \cos(3x)$, $-\pi \le x \le \pi$

264. $f(x) = x^2 e^{-x}$, x is any real number

265. $f(x) = \ln(4 - x^2)$, $-2 < x < 2$

266. Find the absolute maximum and minimum values of $f(x) = x - 2\cos x$ on the interval $[-\pi, \pi]$.

267. Using the following graph of $y = f'(x)$, specify all critical points, assuming the function $y = f(x)$ is defined on $[-4,9]$. Then, specify those critical points at which f attains a local maximum and those at which f attains a local minimum.

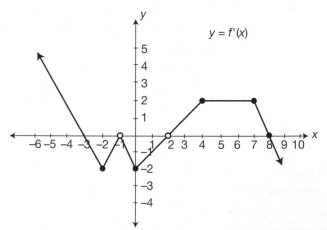

268. Suppose a company makes a profit of $P(x) = \frac{1,000}{x} - \frac{5,000}{x^2} + 100$ dollars when it makes and sells $x > 0$ items. How many items should it make to maximize profit?

269. When 30 orange trees are planted on an acre, each will produce 500 oranges a year. For every additional orange tree planted, each tree will produce 10 oranges less. How many trees should be planted to maximize the yield?

270. An artist can sell 20 copies of a painting at $100 each, but for every copy more that she makes, the value of each painting will go down by a dollar. Thus, if 22 copies are made, each will sell for $98. How many copies should she make to maximize her sales?

271. A garden has 200 pounds of watermelons growing in it. Every day, the total amount of watermelon increases by 5 pounds. At the same time, the price per pound of watermelon goes down by 1¢. If the current price is 90¢ per pound, how much longer should the watermelons grow in order to fetch the highest price possible?

272. A farmer has 400 feet of fencing to make three rectangular pens. What dimensions x and y will maximize the total area?

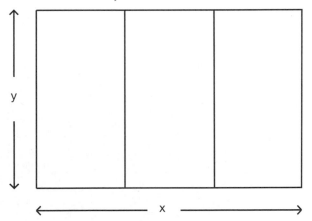

273. Four pens will be built along a river with 150 feet of fencing. What dimensions will maximize the area of the pens?

river (no fence needed)

274. The surface area of a can is Area $= 2\pi r^2 + 2\pi rh$, where the height is h and the radius is r. The volume is Volume $= \pi r^2 h$. What dimensions will minimize the surface area of a can with a volume of 16π cubic inches?

275. A painter has enough paint to cover an area of 600 square feet. What is the largest square-bottomed box that could be painted (including the top, bottom, and all sides)?

Answers

261. Applying the first derivative test requires that we compute the first derivative, as follows:

$f'(x) = 2x^2 + x - 3 = (2x + 3)(x - 1)$

Observe that there are no x-values at which $f'(x)$ is undefined, and the only x-values that make $f'(x) = 0$ are $x = -\frac{3}{2}, 1$. To assess how the sign of $f'(x)$ changes, we form a number line using the critical points, choose a value in each subinterval, and report the sign of $f'(x)$ above the subinterval, as follows:

$f'(x)$ +   −   +

$-\frac{3}{2}$ 1

Since the sign of $f'(x)$ changes from − to + at $x = 1$, and from + to − at $x = -\frac{3}{2}$, we conclude that f has a local minimum at $x = 1$ and a local maximum at $x = -\frac{3}{2}$.

262. Applying the first derivative test requires that we compute the first derivative, as follows:

$f'(x) = \dfrac{(x^2 + 1)(2x) - x^2(2x)}{(x^2 + 1)^2} = \dfrac{2x^3 + 2x - 2x^3}{\left(x^2 + 1\right)^2} = \dfrac{2x}{\left(x^2 + 1\right)^2}$

Observe that there are no x-values at which $f'(x)$ is undefined, and the only x-value that makes $f'(x) = 0$ is $x = 0$. To assess how the sign of $f'(x)$ changes, we form a number line using the critical points, choose a value in each subinterval, and report the sign of $f'(x)$ above the subinterval, as follows:

$f'(x)$ −   +

0

Since the sign of $f'(x)$ changes from − to + at $x = 0$, we conclude that f has a local minimum at $x = 0$. There is no local maximum.

263. Applying the first derivative test requires that we compute the first derivative, as follows:

$f'(x) = -3\sin(3x)$

Observe that there are no x-values at which $f'(x)$ is undefined, and the only x-values that make $f'(x) = 0$ on the given interval $[-\pi, \pi]$ are $x = 0,\ \pm\frac{\pi}{3},\ \pm\frac{2\pi}{3},\ \pm\pi$. To assess how the sign of $f'(x)$ changes, we form a number line using the critical points, choose a value in each subinterval, and report the sign of $f'(x)$ above the subinterval, as follows:

$f'(x)$ $\quad +\quad -\quad +\quad -\quad +\quad -$

$\qquad\qquad -\pi\quad -\frac{2\pi}{3}\quad -\frac{\pi}{3}\quad 0\quad \frac{\pi}{3}\quad \frac{2\pi}{3}\quad \pi$

Since the sign of $f'(x)$ changes from $-$ to $+$ at $x = -\frac{\pi}{3}, \frac{\pi}{3}$, we conclude that f has a local minimum at these values. Likewise, since the sign of $f'(x)$ changes from $+$ to $-$ at $x = -\frac{2\pi}{3}, 0, \frac{2\pi}{3}$, we conclude that f has a local maximum at these values.

264. Applying the first derivative test requires that we compute the first derivative, as follows:

$f'(x) = x^2\left(-e^{-x}\right) + (2x)e^{-x} = xe^{-x}(-x+2)$

Observe that there are no x-values at which $f'(x)$ is undefined, and the only x-values that make $f'(x) = 0$ are $x = 0, 2$. To assess how the sign of $f'(x)$ changes, we form a number line using the critical points, choose a value in each subinterval, and report the sign of $f'(x)$ above the subinterval, as follows:

$f'(x)$ $\quad -\quad +\quad -$

$\qquad\qquad 0\qquad 2$

Since the sign of $f'(x)$ changes from $-$ to $+$ at $x = 0$, and from $+$ to $-$ at $x = 2$, we conclude that f has a local minimum at $x = 0$ and a local maximum at $x = 2$.

265. Applying the first derivative test requires that we compute the first derivative, as follows:

$$f'(x) = \frac{-2x}{4-x^2} = \frac{-2x}{(2-x)(2+x)}$$

Observe that $f'(x)$ is undefined at $x = -2, 2$, and the only x-value that makes $f'(x) = 0$ is $x = 0$. To assess how the sign of $f'(x)$ changes, we form a number line using the critical points, choose a value in each subinterval, and report the sign of $f'(x)$ above the subinterval, as follows:

$$f'(x) \quad \underset{\substack{ \\ -2 \qquad 0 \qquad 2}}{\vdash\!\!\overset{+}{\rule{2.5em}{0pt}}\!\!+\!\!\overset{-}{\rule{2.5em}{0pt}}\!\!\dashv}$$

Since the sign of $f'(x)$ changes from $+$ to $-$ at $x = 0$, we conclude that f has a local maximum at $x = 0$. There is no local minimum.

266. We must identify all critical points within the interval $[-\pi, \pi]$. Among them are the endpoints $\pm\pi$ and x-values at which $f'(x)$ either equals zero or is undefined. Observe that

$$f'(x) = 1 + 2\sin x$$

Observe that there are no x-values at which $f'(x)$ is undefined, and the only x-values in $[-\pi, \pi]$ that make $f'(x) = 0$ are those for which $\sin x = -\frac{1}{2}$, namely $x = -\frac{\pi}{6}, -\frac{5\pi}{6}$. Now, to determine the absolute maximum and minimum values of $f(x) = x - 2\cos x$ on the interval $[-\pi, \pi]$, we simply compute f at each of the critical points, as follows:

$$f(-\pi) = -\pi - 2\cos(-\pi) = -\pi - 2(-1) = 2 - \pi \approx -1.1416$$

$$f\left(-\frac{5\pi}{6}\right) = -\frac{5\pi}{6} - 2\cos\left(-\frac{5\pi}{6}\right) = -\frac{5\pi}{6} - 2\left(-\frac{\sqrt{3}}{2}\right) = -\frac{5\pi}{6} + \sqrt{3} \approx -0.88594$$

$$f\left(-\frac{\pi}{6}\right) = -\frac{\pi}{6} - 2\cos\left(-\frac{\pi}{6}\right) = -\frac{\pi}{6} - 2\left(\frac{\sqrt{3}}{2}\right) = -\frac{\pi}{6} - \sqrt{3} \approx -2.25564$$

$$f(\pi) = \pi - 2\cos(\pi) = \pi - 2(-1) = \pi + 2 \approx 5.1416$$

Hence, the absolute minimum occurs at $x = -\frac{\pi}{6}$ and the absolute maximum occurs at $x = \pi$.

267. Since the function $y = f(x)$ is being considered on $[-4,9]$, the endpoints -4 and 9 are among the critical points. The other critical points consist of those x-values at which $f'(x)$ either is undefined or equals zero. The points at which $f'(x)$ is undefined are those x-values at which there is an open hole in the graph of $y = f'(x)$, namely $x = -1$ and 2. The points at which $f'(x) = 0$ are those for which the graph of $y = f'(x)$ crosses the x-axis and there is not an open hole at the intersection; the x-values for which this is the case are $x = -3$ and 8. Hence, the critical points are: $-4, -3, -1, 2, 8$, and 9.

   Now, to assess how the sign of $f'(x)$ changes, we form a number line using the critical points, choose a value in each subinterval, and report the sign of $f'(x)$ (as determined from the graph) above the subinterval, as follows:

$f'(x)$ +   −   −   +   −

      −4   −3   −1   2   8   9

Since the sign of $f'(x)$ changes from + to − at $x = -3$ and 8, we conclude that f has a local maximum at these values. Similarly, since the sign of $f'(x)$ changes from − to + at $x = 2$, we conclude that f has a local minimum at this value.

268. Applying the first derivative test requires that we compute the first derivative, as follows:

$$P'(x) = -\frac{1,000}{x^2} + \frac{10,000}{x^3}$$

Observe that $P'(x)$ is defined at all positive real numbers, and the only x-value that makes $P'(x) = 0$ is $x = 10$. To assess how the sign of $P'(x)$ changes, we form a number line using the critical points, choose a value in each subinterval, and report the sign of $P'(x)$ above the subinterval, as follows:

$P'(x)$ +   −

      10

Since the sign of $P'(x)$ changes from + to − at $x = 10$, we conclude that P has a local maximum at $x = 10$. Thus, 10 items should be made in order to maximize the profit.

269. If x is the number of trees beyond 30 that are planted on the acre, then the number of oranges produced will be

$$\text{Oranges}(x) = (\text{number of trees}) \cdot (\text{yield per tree})$$
$$= (30 + x)(500 - 10x)$$
$$= 15,000 + 200x - 10x^2$$

Applying the first derivative test requires that we compute the first derivative, as follows:

$$\text{Oranges}'(x) = 200 - 20x$$

Observe that Oranges$'(x)$ is defined at all positive real numbers, and the only x-value that makes it equal zero is $x = 10$. To assess how the sign of Oranges$'(x)$ changes, we form a number line using the critical points, choose a value in each subinterval, and report the sign of Oranges$'(x)$ above the subinterval, as follows:

Oranges$'(x)$ $+$ $-$

          10

Since the sign of Oranges$'(x)$ changes from $+$ to $-$ at $x = 10$, we conclude that Oranges(x) has a local maximum at $x = 10$. Thus, 10 more than 30 trees should be planted, resulting in a total of 40 trees per acre.

270. The total sales will be

$$\text{Sales}(x) = (\text{number of copies}) \cdot (\text{price per copy}) = (20 + x)(100 - x)$$

where x is the number of copies beyond 20. Applying the first derivative test requires that we compute the first derivative, as follows:

$$\text{Sales}'(x) = 80 - 2x$$

Observe that Sales$'(x)$ is for $x \geq 20$, and the only x-value that makes it equal zero is $x = 40$. To assess how the sign of Sales$'(x)$ changes, we form a number line using the critical points, choose a value in each subinterval, and report the sign of Sales$'(x)$ above the subinterval, as follows:

Sales$'(x)$ $+$ $-$

          40

Since the sign of Sales$'(x)$ changes from $+$ to $-$ at $x = 40$, we conclude that Sales(x) has a local maximum at $x = 40$. Thus, the artist should make $x = 40$ more than 20 paintings, for a total of 60 paintings in order to maximize sales.

271. After x days, there will be $200 + 5x$ pounds of watermelons, which will valued at $90 - x$ cents per pound. Thus, the price after x days will be
Price(x) = (pounds of watermelons) × (cents per pound)
$$= (200 + 5x)(90 - x)$$
Applying the first derivative test requires that we compute the first derivative, as follows:

Price$'(x) = 250 - 10x$

Observe that Price$'(x)$ is defined at all real numbers, and the only x-value that makes it equal zero is $x = 25$. To assess how the sign of Price$'(x)$ changes, we form a number line using the critical points, choose a value in each subinterval, and report the sign of Price$'(x)$ above the subinterval, as follows:

Price$'(x)$ $\qquad$ + $\qquad$ –

$\qquad\qquad$ 25

Since the sign of Price$'(x)$ changes from + to – at $x = 25$, we conclude that Price(x) has a local maximum at $x = 25$. Thus, the watermelons will fetch the highest price possible in 25 days.

272. The area is Area $= xy$. Also, by adding the lengths of all sides for which fencing will be used, we see that the total fencing is $4y + 2x = 400$. Thus, $x = 200 - 2y$, so the area function can be written as

Area$(y) = xy = (200 - 2y)y = 200y - 2y^2$

Applying the first derivative test requires that we compute the first derivative, as follows:

Area$'(y) = 200 - 4y$

Observe that Area$'(y)$ is defined at all real numbers, and the only y-value that makes it equal zero is $y = 50$. To assess how the sign of Area$'(y)$ changes, we form a number line using the critical points, choose a value in each subinterval, and report the sign of Area$'(y)$ above the subinterval, as follows:

Area$'(y)$ $\qquad$ + $\qquad$ –

$\qquad\qquad$ 50

Since the sign of Area$'(y)$ changes from + to – at $y = 50$, we conclude that Area(y) has a local maximum at $y = 50$. Thus, the optimal dimensions for each pen are $y = 50$ feet and $x = 200 - 2y = 200 - 2(50)$ $= 100$ feet.

273. The area is Area $= xy$. Also, summing the lengths of all sides for which fencing will be used, we see that the total fencing is $5y + x = 150$. Thus, $x = 150 - 5y$, so the area function can be written as

$$\text{Area}(y) = xy = (150 - 5y)y = 150y - 5y^2$$

Applying the first derivative test requires that we compute the first derivative, as follows:

$$\text{Area}'(y) = 150 - 10y$$

Observe that Area$'(y)$ is defined at all real numbers, and the only y-value that makes it equal zero is $y = 15$. To assess how the sign of Area$'(y)$ changes, we form a number line using the critical points, choose a value in each subinterval, and report the sign of Area$'(y)$ above the subinterval, as follows:

Area$'(y)$ $+$ $-$

                    15

Since the sign of Area$'(y)$ changes from $+$ to $-$ at $y = 15$, we conclude that Area(y) has a local maximum at $y = 15$. Thus, the optimal dimensions of each pen are $y = 15$ feet and $x = 150 - 5(15) = 75$ feet.

274. Since Volume $= \pi r^2 h = 16\pi$, it follows that $h = \frac{16}{r^2}$. Thus, the surface area function is given by

$$\text{Area}(r) = 2\pi r^2 + 2\pi r \left(\frac{16}{r^2} \right) = 2\pi r^2 + \frac{32\pi}{r}$$

Applying the first derivative test requires that we compute the first derivative, as follows:

$$\text{Area}'(r) = 4\pi r - \frac{32\pi}{r^2}$$

Observe that Area$'(r)$ is defined at all nonzero real numbers, and the only r-value that makes it equal zero is when $4\pi r = \frac{32\pi}{r^2}$, so that $r^3 = 8$, or equivalently $r = 2$. To assess how the sign of Area$'(r)$ changes, we form a number line using the critical points, choose a value in each subinterval, and report the sign of Area$'(r)$ above the subinterval, as follows:

Area$'(r)$ $-$ $+$

$\longleftarrow\!\!\!\!\!\!\underset{2}{\rule{0pt}{1em}\big|}\!\!\!\!\!\!\longrightarrow$

Since the sign of Area$'(r)$ changes from $-$ to $+$ at $r = 2$, we conclude that Area(r) has a local minimum at $r = 2$. Thus, a radius of $r = 2$ inches and a height of $h = \frac{16}{r^2} = 4$ inches will minimize the surface area.

275. Since the box has a square bottom, its length and width can be both x, while its height is y. Thus the volume is given by Volume $= x^2 y$ and the surface area is Area $= x^2 + 4xy + x^2$ (the top, the four sides, and the bottom). Since Area $= 2x^2 + 4xy = 600$, it follows that the height $y = \frac{600 - 2x^2}{4x} = \frac{150}{x} - \frac{x}{2}$. Thus, the volume function is given by

$$\text{Volume}(x) = x^2 y = x^2 \left(\frac{150}{x} - \frac{x}{2} \right) = 150x - \frac{1}{2}x^3$$

Applying the first derivative test requires that we compute the first derivative, as follows:

$$\text{Volume}'(x) = 150 - \frac{3}{2}x^2$$

Observe that Volume$'(x)$ is defined at all nonzero real numbers, and the only x-value that makes it equal zero is when $x^2 = 100$. Since negative lengths are impossible, this is only zero when $x = 10$. To assess how the sign of Volume$'(x)$ changes, we form a number line using the critical points, choose a value in each subinterval, and report the sign of Volume$'(x)$ above the subinterval, as follows:

Volume$'(x)$ $+$ $-$

               10

Since the sign of Volume$'(x)$ changes from $+$ to $-$ at $x = 10$, we conclude that Volume(x) has a local maximum at $x = 10$ feet. The corresponding height is $y = \frac{150}{10} - \frac{10}{2} = 10$ feet, so we conclude that the largest box that could be painted is a cube with all sides of length 10 feet.

The Integral: Definition, Properties, and Fundamental Theorem of Calculus Problems

The original and most common interpretation of the integral $\int_a^b f(x)\,dx$ is linked to the area of the region bounded by the curve $y = f(x)$, the x-axis, and the lines $x = a$ and $x = b$. If the curve lies above the x-axis, then the integral *is* this area, whereas if the curve lies below the x-axis, the integral is -1 times the area of the region. The following are some useful properties of the integral.

Properties of the Integral

Let k be a real number and

1. $\int_a^b f(x)\,dx = -\int_b^a f(x)\,dx$

2. **Linearity:** $\int_a^b kf(x)\,dx = k\int_a^b f(x)\,dx$,

    $\int_a^b \left[f(x) \pm g(x)\right]dx = \int_a^b f(x)\,dx \pm \int_a^b g(x)\,dx$

3. **Interval additivity:** $\int_a^b f(x)\,dx + \int_b^c f(x)\,dx = \int_a^c f(x)\,dx$

4. **Symmetry: i.** If f is an even function on $[-a,a]$, then
    $$\int_{-a}^a f(x)\,dx = 2\int_0^a f(x)\,dx.$$

    ii. If f is an odd function on $[-a,a]$, then $\int_{-a}^a f(x)\,dx = 0$.

5. **Periodicity:** If f has period p, then $\int_{a+np}^{b+np} f(x)\,dx = \int_a^b f(x)\,dx$, for any integer n.

6. **Antiderivative rule:** $\int_a^b g'(x)\,dx = g(x)\big|_a^b = g(b) - g(a)$

7. **Fundamental theorem of calculus:** If f is continuous on $[a,b]$, then $\frac{d}{dx}\int_a^x f(t)\,dt = f(x)$. More generally, if u is a differentiable function, then

    the chain rule yields $\frac{d}{dx}\int_a^{u(x)} f(t)\,dt = f(u(x)) \cdot u'(x)$.

Questions

For Questions 276 through 278, use the following graph.

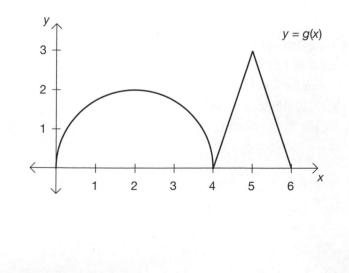

276. $\displaystyle\int_0^4 g(x)\,dx$

277. $\displaystyle\int_4^6 g(x)\,dx$

278. $\displaystyle\int_0^6 g(x)\,dx$

For Questions 279 through 281, use the following graph.

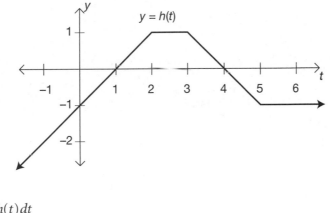

279. $\displaystyle\int_{-1}^6 h(t)\,dt$

280. $\displaystyle\int_{-1}^4 h(t)\,dt$

281. $\displaystyle\int_4^6 h(t)\,dt$

For Questions 282 through 284, use the following graph.

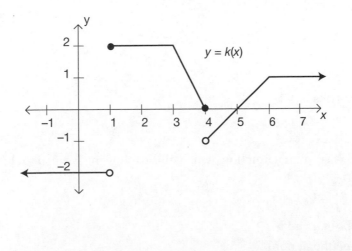

282. $\int_0^7 k(x)\,dx$

283. $\int_4^6 k(x)\,dx$

284. $\int_4^5 k(x)\,dx$

For Questions 285 and 286, assume that $\int_0^6 f(x)\,dx = 10$, $\int_6^7 f(x)\,dx = -5$, and $\int_7^{11} f(x)\,dx = 2$.

285. $\int_0^7 5f(x)\,dx$

286. $\int_6^{11} -2f(x)\,dx$

For Questions 287 through 289, assume that $\int_1^{14} g(t)\,dt = -3$, $\int_{10}^{14} g(t)\,dt = 8$, and $\int_1^5 g(t)\,dt = -10$.

287. $\int_5^{14} 3g(t)\,dt$

288. $\int_1^{10} -4g(t)\,dt$

289. $\int_5^{10} 2g(t)\,dt$

290. True or false? $\int_0^{\frac{\pi}{2}} \sin x\,dx = \int_{2n\pi}^{\frac{\pi}{2}+2n\pi} \sin x\,dx$, for any integer n.

291. If $f(x)$ is an odd function, which of the following is equivalent to $\int_{-1}^2 f(x)\,dx$?

    a. $-\int_{-1}^2 f(-x)\,dx$

    b. $\int_2^{-1} f(-x)\,dx$

    c. both **a** and **b**

    d. neither **a** nor **b**

292. If $f(x)$ is an even function, which of the following is equal to $\int_{-5}^5 f^2(x)\,dx$?

    a. 0

    b. $2\int_0^5 f^2(x)\,dx$

    c. both **a** and **b**

    d. neither **a** nor **b**

293. True or false? If the range of g is $(-5,-1)$ and its domain is the set of all real numbers, $\mathbb{R}$, then $\int_0^{-1} g(x)\,dx < 0$.

294. Which of the following is equivalent to $\int_{-\frac{\pi}{4}}^{\frac{\pi}{4}} \tan(\frac{1}{2}x)\,dx$?

   a. $\int_{\frac{7\pi}{4}}^{\frac{9\pi}{4}} \tan(\frac{1}{2}x)\,dx$

   b. $\int_{\frac{\pi}{4}}^{\frac{3\pi}{4}} \tan(\frac{1}{2}x)\,dx$

   c. both **a** and **b**

   d. neither **a** nor **b**

295. Which of the following is equivalent to $\frac{d}{dx}\int_{-2}^{x} \sqrt{t^2+1}\,dt$?

   a. $-\sqrt{x^2+1}$

   b. $\sqrt{x^2+1}-\sqrt{5}$

   c. $\sqrt{x^2+1}$

   d. $\sqrt{5}-\sqrt{x^2+1}$

296. Which of the following is equivalent to $\frac{d}{dx}\int_3^{6x^2+2} \sin|t-1|\,dt$?

   a. $\sin|6x^2+1|$

   b. $\sin|6x^2+1|-\sin 2$

   c. $12x\sin|6x^2+1|$

   d. $12x\sin|6x^2+1|-12(\sin 2)x$

297. Which of the following is equivalent to $\frac{d}{dx}\int_{xe^x}^{1} \ln\left(2\sqrt[3]{t}+1\right)dt$?

   a. $-e^x(x+1)\ln\left(2\sqrt[3]{xe^x}+1\right)$

   b. $e^x(x+1)\ln\left(2\sqrt[3]{xe^x}+1\right)$

   c. $-e^x(x+1)\ln\left(2\sqrt[3]{xe^x}+1\right)+\ln 3$

   d. $e^x(x+1)\ln\left(2\sqrt[3]{xe^x}+1\right)-\ln 3$

298. Which of the following is equivalent to $\frac{d}{dx}\int_{\ln x}^{2}\ln\left(te^{t}\right)dt$?

a. $-\dfrac{\ln(\ln x+x)}{x}$

b. $\dfrac{\ln(\ln x+x)}{x}$

c. $\dfrac{\ln(\ln x)+\ln x}{x}$

d. $-\dfrac{\ln(\ln x)+\ln x}{x}$

299. Which of the following is equivalent to $\int_{\tan x}^{\cot x}\dfrac{1}{t^{2}+1}dt$?

a. $\dfrac{1}{\cot^{2}x+1}-\dfrac{1}{\tan^{2}x+1}$

b. $\dfrac{-\csc^{2}x}{\cot^{2}x+1}$

c. $\dfrac{1}{\cot^{2}x+1}$

d. -2

300. Which of the following is equivalent to $\frac{d}{dx}\int_{x}^{-1}\left(4t^{4}+2\right)^{3}dt$?

a. $-\left(4x^{4}+2\right)^{3}$

b. $\left(4x^{4}+2\right)^{3}$

c. $-\left(4x^{4}+2\right)^{3}\cdot 3\left(4x^{4}+2\right)^{2}\left(16x^{3}\right)$

d. $\left(4x^{4}+2\right)^{3}\cdot 3\left(4x^{4}+2\right)^{2}\left(16x^{3}\right)$

Answers

276. $\int_0^4 g(x)dx$ = the area of a semicircle with radius 2. This area is
$\frac{1}{2}\pi(2)^2 = 2\pi$.

277. $\int_4^6 g(x)dx$ = the area of a triangle with base length 2 and height 3. This
area is $\frac{1}{2}(2)(3)=3$.

278. Using interval additivity, $\int_0^6 g(x)dx = \int_0^4 g(x)dx + \int_4^6 g(x)dx = 2\pi+3$.

279. We break the entire region into discernible geometric shapes so that
known area formulas can be applied. Then, using interval additivity
yields
$$\int_{-1}^6 h(t)dt = \int_{-1}^1 h(t)dt + \int_1^4 h(t)dt + \int_4^5 h(t)dt + \int_5^6 h(t)dt$$
For any region that lies below the x-axis, we determine the area of the
region and multiply it by −1 to obtain the value of the integral; if the
region lies entirely above the x-axis, the area of the region is equal to the
value of the integral. Doing so yields

$\int_{-1}^1 h(t)dt$ = −1 times the area of a triangle with base 2 and height 2,

namely $-\frac{1}{2}(2)(2)=-2$.

$\int_1^4 h(t)dt$ = the area of a trapezoid with bases 1 and 3 and height 1,

namely $\frac{1}{2}(1)(1+3)=2$.

$\int_4^5 h(t)dt$ = −1 times the area of a triangle with base 1 and height 1,

namely $-\frac{1}{2}(1)(1)=-\frac{1}{2}$.

$\int_5^6 h(t)dt$ = −1 times the area of a square with side 1, namely
$-(1)(1)=-1$.

Now, substituting these values into the expression yields

$\int_{-1}^6 h(t)dt = -2+2-\frac{1}{2}-1=-\frac{3}{2}$

280. We break the entire region into discernible geometric shapes so that known area formulas can be applied. Then, using interval additivity yields

$$\int_{-1}^{4} h(t)\,dt = \int_{-1}^{1} h(t)\,dt + \int_{1}^{4} h(t)\,dt$$

For any region that lies below the x-axis, we determine the area of the region and multiply it by -1 to obtain the value of the integral; if the region lies entirely above the x-axis, the area of the region is equal to the value of the integral. Doing so yields

$\int_{-1}^{1} h(t)\,dt = -1$ times the area of a triangle with base 2 and height 2,

namely $-\frac{1}{2}(2)(2) = -2$.

$\int_{1}^{4} h(t)\,dt =$ the area of a trapezoid with bases 1 and 3 and height 1,

namely $\frac{1}{2}(1)(1+3) = 2$.

Now, substituting these values into the expression yields

$$\int_{-1}^{4} h(t)\,dt = -2 + 2 = 0$$

281. We break the entire region into discernible geometric shapes so that known area formulas can be applied. Then, using interval additivity yields

$$\int_{4}^{6} h(t)\,dt = \int_{4}^{5} h(t)\,dt + \int_{5}^{6} h(t)\,dt$$

For any region that lies below the x-axis, we determine the area of the region and multiply it by -1 to obtain the value of the integral; if the region lies entirely above the x-axis, the area of the region is equal to the value of the integral. Doing so yields

$\int_{4}^{5} h(t)\,dt = -1$ times the area of a triangle with base 1 and height 1,

namely $-\frac{1}{2}(1)(1) = -\frac{1}{2}$.

$\int_{5}^{6} h(t)\,dt = -1$ times the area of a square with side 1, namely
$-(1)(1) = -1$.

Now, substituting these values into the expression yields

$$\int_{4}^{6} h(t)\,dt = -\frac{1}{2} - 1 = -\frac{3}{2}$$

282. We break the entire region into discernible geometric shapes so that known area formulas can be applied. Then, using interval additivity yields

$$\int_0^7 k(x)\,dx = \int_0^1 k(x)\,dx + \int_1^3 k(x)\,dx + \int_3^4 k(x)\,dx + \int_4^5 k(x)\,dx + \int_5^6 k(x)\,dx$$

$$+ \int_6^7 k(x)\,dx$$

For any region that lies below the x-axis, we determine the area of the region and multiply it by -1 to obtain the value of the integral; if the region lies entirely above the x-axis, the area of the region is equal to the value of the integral. Doing so yields

$\int_0^1 k(x)\,dx = -1$ times the area of a rectangle with sides 1 and 2,

namely $-(2)(1) = -2$.

$\int_1^3 k(x)\,dx =$ the area of a rectangle with sides 2 and 2, namely $2(2) = 4$.

$\int_3^4 k(x)\,dx =$ the area of a rectangle with base 1 and height 2, namely

$\frac{1}{2}(2)(1) = 1$.

$\int_4^5 k(x)\,dx = -1$ times the area of a triangle with base 1 and height 1,

namely $-\frac{1}{2}(1)(1) = -\frac{1}{2}$.

$\int_5^6 k(x)\,dx =$ the area of a triangle with base 1 and height 1, namely

$\frac{1}{2}(1)(1) = \frac{1}{2}$.

$\int_6^7 k(x)\,dx =$ the area of a square with side 1, namely $1(1) = 1$.

Now, substituting these values into the expression yields

$$\int_0^7 k(x)\,dx = -2 + 4 + 1 - \frac{1}{2} + \frac{1}{2} + 1 = 4$$

283. We break the entire region into discernible geometric shapes so that known area formulas can be applied. Then, using interval additivity yields

$$\int_4^6 k(x)\,dx = \int_4^5 k(x)\,dx + \int_5^6 k(x)\,dx$$

For any region that lies below the x-axis, we determine the area of the region and multiply it by −1 to obtain the value of the integral; if the region lies entirely above the x-axis, the area of the region is equal to the value of the integral. Doing so yields

$$\int_4^5 k(x)\,dx = -1 \text{ times the area of a triangle with base 1 and height 1,}$$

namely $-\frac{1}{2}(1)(1) = -\frac{1}{2}$.

$$\int_5^6 k(x)\,dx = \text{ the area of a triangle with base 1 and height 1, namely}$$

$$\frac{1}{2}(1)(1) = \frac{1}{2}.$$

Now, substituting these values into the expression yields

$$\int_4^6 k(x)\,dx = -\frac{1}{2} + \frac{1}{2} = 0$$

284. Observe that the region lies below the x-axis. So, we determine the area of the region and multiply it by −1 to obtain the value of the integral.

$$\int_4^5 k(x)\,dx = -1 \text{ times the area of a triangle with base 1 and height 1,}$$
namely $-\frac{1}{2}(1)(1) = -\frac{1}{2}$.

285. Using interval additivity and linearity, we see that

$$\int_0^7 5f(x)\,dx = 5\int_0^7 f(x)\,dx = 5\left[\int_0^6 f(x)\,dx + \int_6^7 f(x)\,dx\right] = 5[10+(-5)] = 25$$

286. Using interval additivity and linearity, we see that

$$\int_6^{11} -2f(x)\,dx = -2\int_6^{11} f(x)\,dx = -2\left[\int_6^7 f(x)\,dx + \int_7^{11} f(x)\,dx\right]$$
$$= -2[-5+2] = 6$$

287. Using interval additivity and linearity, we see that

$$\int_5^{14} 3g(t)\,dt = 3\int_5^{14} g(t)\,dt = 3\left[\int_1^{14} g(t)\,dt - \int_1^5 g(t)\,dt\right] = 3[-3-(-10)] = 21$$

288. Using interval additivity and linearity, we see that

$$\int_1^{10} -4g(t)\,dt = -4\int_1^{10} g(t)\,dt = -4\left[\int_1^{14} g(t)\,dt - \int_{10}^{14} g(t)\,dt\right] = -4[-3-8]$$
$$= 44$$

289. Using interval additivity and linearity, we see that

$$\int_5^{10} 2g(t)\,dt = 2\int_5^{10} g(t)\,dt = 2\left[\int_1^{14} g(t)\,dt - \int_{10}^{14} g(t)\,dt - \int_1^{5} g(t)\,dt\right]$$

$$= 2[-3-8-(-10)] = -2$$

290. True. Since the period of $f(x) = \sin x$ is 2π, we conclude that these two integrals are equal using the periodicity property with $p = 2\pi$.

291. **c.** Since $f(x)$ is an odd function, we know that $f(x) = -f(-x)$, for any x in the domain of f. Hence, $\int_{-1}^{2} f(x)\,dx = \int_{-1}^{2} -f(-x)\,dx$. Using linearity, the integral on the right side is equal to $-\int_{-1}^{2} f(-x)\,dx$. Moreover, using property 1 from the beginning of the chapter, this integral is equal to $\int_{2}^{-1} f(-x)\,dx$.

292. **b.** Since $f(x)$ is an even function, then $f(x) = f(-x)$. Thus, $f^2(x) = f^2(-x)$, so $f^2(x)$ is also an even function. Hence, property $4(i)$ implies that $\int_{-5}^{5} f^2(x)\,dx = 2\int_{0}^{5} f^2(x)\,dx$.

293. False. Since the range of g is $(-5,-1)$, we know that $g(x) < 0$ for all $-1 < x < 0$. Hence, $\int_{-1}^{0} g(x)\,dx < 0$. Since $\int_{0}^{-1} g(x)\,dx = -\int_{-1}^{0} g(x)\,dx$, it follows that $\int_{0}^{-1} g(x)\,dx > 0$.

294. **a.** The period of $f(x) = \tan\left(\frac{1}{2}x\right)$ is 2π. So, using property 5 with $n = 1$ yields the equality

$$\int_{-\frac{\pi}{4}}^{\frac{\pi}{4}} \tan(\tfrac{1}{2}x)\,dx = \int_{-\frac{\pi}{4}+2\pi}^{\frac{\pi}{4}+2\pi} \tan(\tfrac{1}{2}x)\,dx = \int_{\frac{7\pi}{4}}^{\frac{9\pi}{4}} \tan(\tfrac{1}{2}x)\,dx$$

Also note that the integral in choice **b** is not equivalent to the original integral because only $\frac{\pi}{2}$ is being added to both limits, and this is not the period of the integrand.

295. **c.** Using the fundamental theorem of calculus immediately yields

$$\frac{d}{dx}\int_{-2}^{x} \sqrt{t^2 + 1}\,dt = \sqrt{x^2 + 1}$$

296. **c.** Using the more general form of the fundamental theorem of calculus involving the chain rule yields

$$\frac{d}{dx}\int_{3}^{6x^2+2} \sin|t-1|\,dt = \sin|6x^2 + 1| \cdot \frac{d}{dx}\left(6x^2 + 1\right) = 12x\sin|6x^2 + 1|$$

297. a. Before applying the fundamental theorem of calculus, we must apply property 1 in order to get the integral into the proper form to which the rule applies. This yields

$$\frac{d}{dx}\int_{xe^x}^{1}\ln\left(2\sqrt[3]{t}+1\right)dt = \frac{d}{dx}\left[-\int_{1}^{xe^x}\ln\left(2\sqrt[3]{t}+1\right)dt\right] = -\frac{d}{dx}\int_{1}^{xe^x}\ln\left(2\sqrt[3]{t}+1\right)dt$$

Now, applying the more general form of the fundamental theorem of calculus involving the chain rule yields

$$\frac{d}{dx}\int_{xe^x}^{1}\ln\left(2\sqrt[3]{t}+1\right)dt = -\frac{d}{dx}\int_{1}^{xe^x}\ln\left(2\sqrt[3]{t}+1\right)dt$$

$$= -\ln\left(2\sqrt[3]{xe^x}+1\right)\cdot\frac{d}{dx}\left(xe^x\right)$$

$$= -e^x(x+1)\ln\left(2\sqrt[3]{xe^x}+1\right)$$

298. d. Before applying the fundamental theorem of calculus, we must apply property 1 in order to get the integral into the proper form to which the rule applies. This yields

$$\frac{d}{dx}\int_{\ln x}^{2}\ln(te^t)\,dt = \frac{d}{dx}\left[-\int_{2}^{\ln x}\ln(te^t)\,dt\right] = -\frac{d}{dx}\int_{2}^{\ln x}\ln(te^t)\,dt$$

Now, applying the more general form of the fundamental theorem of calculus involving the chain rule yields

$$\frac{d}{dx}\int_{\ln x}^{2}\ln(te^t)\,dt = -\frac{d}{dx}\int_{2}^{\ln x}\ln(te^t)\,dt$$

$$= -\ln\left(\ln x\cdot e^{\ln x}\right)\cdot\frac{d}{dx}(\ln x)$$

$$= -\left[\ln(\ln x)+\ln\left(e^{\ln x}\right)\right]\cdot\frac{d}{dx}(\ln x)$$

$$= -\left[\ln(\ln x)+\ln(x)\right]\cdot\left(\frac{1}{x}\right)$$

$$= -\frac{\ln(\ln x)+\ln x}{x}$$

(Note that the logarithm rules were used in the third and fourth lines of the preceding string of equalities.)

501 Calculus Questions

299. d. First, we must apply interval additivity to express the given integral as a sum of two integrals, each of which has at least one limit that is constant. Using interval additivity and then property 1 yields

$$\int_{\tan x}^{\cot x} \frac{1}{t^2+1}\,dt = \int_{\tan x}^{0} \frac{1}{t^2+1}\,dt + \int_{0}^{\cot x} \frac{1}{t^2+1}\,dt = -\int_{0}^{\tan x} \frac{1}{t^2+1}\,dt + \int_{0}^{\cot x} \frac{1}{t^2+1}\,dt$$

Now, applying the more general form of the fundamental theorem of calculus involving the chain rule yields

$$\frac{d}{dx}\int_{\tan x}^{\cot x} \frac{1}{t^2+1}\,dt = \frac{d}{dx}\left[-\int_{0}^{\tan x} \frac{1}{t^2+1}\,dt + \int_{0}^{\cot x} \frac{1}{t^2+1}\,dt\right]$$

$$= -\frac{d}{dx}\int_{0}^{\tan x} \frac{1}{t^2+1}\,dt + \frac{d}{dx}\int_{0}^{\cot x} \frac{1}{t^2+1}\,dt$$

$$= -\frac{1}{\tan^2 x+1}\cdot\frac{d}{dx}(\tan x) + \frac{1}{\cot^2 x+1}\cdot\frac{d}{dx}(\cot x)$$

$$= -\frac{\sec^2 x}{\tan^2 x+1} + \frac{-\csc^2 x}{\cot^2 x+1}$$

$$= -\frac{\sec^2 x}{\sec^2 x} - \frac{\csc^2 x}{\csc^2 x}$$

$$= -1-1$$

$$= -2$$

300. a. Before applying the fundamental theorem of calculus, we must apply property 1 in order to get the integral into the proper form to which the rule applies. This yields

$$\frac{d}{dx}\int_{x}^{-1}\left(4t^4+2\right)^3 dt = \frac{d}{dx}\left[-\int_{1}^{x}\left(4t^4+2\right)^3 dt\right] = -\frac{d}{dx}\int_{1}^{x}\left(4t^4+2\right)^3 dt$$

Now, applying the more general form of the fundamental theorem of calculus involving the chain rule yields

$$\frac{d}{dx}\int_{x}^{-1}\left(4t^4+2\right)^3 dt = -\frac{d}{dx}\int_{1}^{x}\left(4t^4+2\right)^3 dt = -\left(4x^4+2\right)^3$$

Integration Techniques Problems

The so-called *second fundamental theorem of calculus* links the process of integration to finding an antiderivative, as follows:

Second Fundamental Theorem of Calculus: If f is a continuous function and F is an antiderivative of f—that is, $F'(x) = f(x)$—then $\int_a^b f(x)\,dx = F(x)\big|_a^b = F(b) - F(a)$.

The evaluation symbol $F(x)\big|_a^b$ is just a way of keeping track of the *limits of integration a* and *b* before they are plugged into $F(x)$ and subtracted. Applying this theorem relies on our ability to determine an antiderivative of the integrand. The simplest integrals to apply this theorem to are those for which an antiderivative is known.

Some Common Antiderivatives

The following is a list of known antiderivatives, where C denotes an arbitrary constant. When an integral is listed without limits, it is referred to as an *indefinite integral*.

1. $\int \sin x\, dx = -\cos x + C$

2. $\int \cos x\, dx = \sin x + C$

3. $\int \sec^2 x\, dx = \tan x + C$

4. $\int \csc^2 x\, dx = -\cot x + C$

5. $\int \sec x \tan x\, dx = \sec x + C$

6. $\int \csc x \cot x\, dx = -\csc x + C$

7. $\int e^x\, dx = e^x + C$

8. $\int \frac{1}{x}\, dx = \ln|x| + C$

9. $\int x^n\, dx = \frac{x^{n+1}}{n+1} + C,\ n \neq -1$

10. $\int \frac{1}{1+x^2}\, dx = \arctan x + C$

These formulas can be used in conjunction with a variety of techniques, including substitution, integration by parts, trigonometric integrals, and partial fraction decomposition (and combinations thereof) to compute the integrals of more complicated functions. This section focuses on computing indefinite integrals using these techniques.

Questions

For Questions 301 through 310, compute the indefinite integral.

301. $\int 5x^{12}\, dx$

302. $\int \frac{1}{\sqrt[4]{x}}\, dx$

303. $\int \left(x - 2x^3 \sqrt{x}\right) dx$

304. $\int \left(2x^{-3} - 3x^{-2} - 4x^{-1} + e \right) dx$

305. $\int \left(-e^x + 2x^\pi \right) dx$

306. $\int \left(-x + \pi \cos x \right) dx$

307. $\int \left(-\pi + \frac{x^{\frac{2}{5}} \sqrt[4]{x}}{x^{-4}} \right) dx$

308. $\int \left(-\sec^2 x + 3 \cot x \csc x \right) dx$

309. $\int \left(-\frac{1}{1+x^2} - \frac{1}{x^2} + \left(\frac{1}{x^{-2}} \right)^{-2} \right) dx$

310. $\int \left(2\sec x \tan x - 3e^x + 4\csc^2 x \right) dx$

For Questions 311 through 324, use the method of u-substitution to compute the indefinite integral.

311. $\int x^3 \left(1 - x^4 \right)^5 dx$

312. $\int 2x \sqrt{4x^2 - 3} \, dx$

313. $\int \frac{x - x^4}{4x^5 - 10x^2 + 3} dx$

314. $\int \cos^2 x \sin x \, dx$

315. $\int x^2 e^{3x^3 + 1} \, dx$

316. $\int \frac{e^{3x}}{2 - e^{3x}} dx$

317. $\int \frac{\tan \left(\sqrt{x} + 1 \right) \sec^2 \left(\sqrt{x} + 1 \right)}{\sqrt{x}} dx$

318. $\int \frac{3x}{1 + 4x^4} dx$

319. $\int \frac{e^x}{1 - \sin^2 \left(e^x \right)} dx$

320. $\int \cot (\pi x) dx$

321. $\int \dfrac{\sqrt[5]{\ln(x+1)}}{x+1}\,dx$

322. $\int \dfrac{2}{e^{\sqrt[3]{x}}\cdot\sqrt[3]{x^2}}\,dx$

323. $\int \dfrac{e^{3x+6}\left(e^x\right)^3}{e^{2x+1}}\,dx$

324. $\int \dfrac{x^2\cos\left(\dfrac{1}{x^3+8}\right)}{\left(x^3+8\right)^2}\,dx$

For Questions 325 through 330, use integration by parts to compute the indefinite integral.

325. $\int x\cos(3x)\,dx$

326. $\int \left(x^2+2x+5\right)e^{-2x}\,dx$

327. $\int \cos(2x)\cdot\ln\left(\sin(2x)\right)\,dx$

328. $\int e^{-x}\sin x\,dx$

329. $\int \dfrac{3\ln\left(\ln\left(e^x+1\right)\right)e^x}{e^x+1}\,dx$

330. $\int \arctan(4x)\,dx$

For Questions 331 through 334, use partial fraction decomposition to aid in the computation of the indefinite integral.

331. $\int \dfrac{1}{x(x-1)}\,dx$

332. $\int \dfrac{9x-2}{(x-1)^2}\,dx$

333. $\int \dfrac{9x-11}{(x-3)(x+5)}\,dx$

334. $\int \dfrac{x^3}{\left(x^2+9\right)^2}\,dx$

For Questions 335 through 340, compute the indefinite integral using an appropriate technique.

335. $\displaystyle\int \sin^3\left(\frac{\pi}{2}x\right)dx$

336. $\displaystyle\int \sin^2(3x)dx$

337. $\displaystyle\int \sin^2 x \cos^3 x\,dx$

338. $\displaystyle\int \frac{2}{3+7x^2}dx$

339. $\displaystyle\int \frac{\left(x^3-1\right)^2\left(x^3+1\right)^2}{x^6}dx$

340. $\displaystyle\int \frac{1+\ln x}{\sin^2(x\ln x)}dx$

Answers

301. $\int 5x^{12}\,dx = 5\int x^{12}\,dx = 5\cdot\frac{x^{13}}{13}+C = \frac{5}{13}x^{13}+C$

302. $\int \frac{1}{\sqrt[4]{x}}\,dx = \int \frac{1}{x^{\frac{1}{4}}}\,dx = \int x^{-\frac{1}{4}}\,dx = \frac{x^{\frac{3}{4}}}{\frac{3}{4}}+C = \frac{4}{3}x^{\frac{3}{4}}+C$

303. $\int\left(x-2x^{3}\sqrt{x}\right)dx = \int\left(x-2x^{\frac{7}{2}}\right)dx = \frac{x^{2}}{2}-2\cdot\frac{x^{\frac{9}{2}}}{\frac{9}{2}}+C = \frac{1}{2}x^{2}-\frac{4}{9}x^{\frac{9}{2}}+C$

304. $\int\left(2x^{-3}-3x^{-2}-4x^{-1}+e\right)dx = 2\cdot\frac{x^{-2}}{-2}-3\cdot\frac{x^{-1}}{-1}-4\ln|x|+ex+C$

$$= -\frac{1}{x^{2}}+\frac{3}{x}-4\ln|x|+ex+C$$

305. $\int\left(-e^{x}+2x^{\pi}\right)dx = -e^{x}+2\cdot\frac{x^{\pi+1}}{\pi+1}+C$

306. $\int\left(-x+\pi\cos x\right)dx = -\frac{x^{2}}{2}+\pi\sin x+C$

307. Using the exponent rules, we see that $\dfrac{x^{\frac{2}{5}}\sqrt[4]{x}}{x^{-4}} = \dfrac{x^{\frac{2}{5}}x^{\frac{1}{4}}}{x^{-4}} = x^{4+\frac{2}{5}+\frac{1}{4}} = x^{\frac{93}{20}}$.
Thus,

$$\int\left(-\pi+\frac{x^{\frac{2}{5}}\sqrt[4]{x}}{x^{-4}}\right)dx = \int\left(-\pi+x^{\frac{93}{20}}\right)dx = -\pi x+\frac{x^{\frac{113}{20}}}{\frac{113}{20}}+C = -\pi x+\frac{20}{113}x^{\frac{113}{20}}+C$$

308. $\int\left(-\sec^{2}x+3\cot x\csc x\right)dx = -\tan x-3\csc x+C$

309. $\int\left(-\frac{1}{1+x^{2}}-\frac{1}{x^{2}}+\left(\frac{1}{x^{-2}}\right)^{-2}\right)dx = \int\left(-\frac{1}{1+x^{2}}-x^{-2}+x^{-4}\right)dx$

$$= -\arctan x-\frac{x^{-1}}{-1}+\frac{x^{-3}}{-3}+C$$

$$= -\arctan x+\frac{1}{x}-\frac{1}{3x^{3}}+C$$

310. $\int\left(2\sec x\tan x-3e^{x}+4\csc^{2}x\right)dx = 2\sec x-3e^{x}-4\cot x+C$

311. Make the following substitution:

$u = 1 - x^4$

$du = -4x^3 dx \implies -\frac{1}{4} du = x^3 dx$

Applying this substitution in the integrand and computing the resulting indefinite integral yields

$$\int x^3 \left(1 - x^4\right)^5 dx = \int \left(1 - x^4\right)^5 x^3 dx = \int u^5 \cdot \left(-\frac{1}{4}\right) du = -\frac{1}{4} \int u^5 du$$

$$= -\frac{1}{4} \cdot \frac{u^6}{6} + C$$

Finally, rewrite the final expression of the preceding equation in terms of the original variable x by resubstituting $u = 1 - x^4$ to obtain

$$\int x^3 \left(1 - x^4\right)^5 dx = -\frac{1}{4} \cdot \frac{\left(1 - x^4\right)^6}{6} + C = -\frac{\left(1 - x^4\right)^6}{24} + C$$

312. Make the following substitution:

$u = 4x^2 - 3$

$du = 8x dx \implies \frac{1}{8} du = x dx$

Applying this substitution in the integrand and computing the resulting indefinite integral yields

$$\int 2x \sqrt{4x^2 - 3}\, dx = 2 \int \sqrt{4x^2 - 3}\, x\, dx = 2 \int \sqrt{u} \left(\frac{1}{8}\right) du = \frac{1}{4} \int u^{\frac{1}{2}} du$$

$$= \frac{1}{4} \cdot \frac{u^{\frac{3}{2}}}{\frac{3}{2}} + C = \frac{1}{6} u^{\frac{3}{2}} + C$$

Finally, rewrite the final expression of the preceding equation in terms of the original variable x by resubstituting $u = 4x^2 - 3$ to obtain

$$\int 2x \sqrt{4x^2 - 3}\, dx = \frac{1}{6}\left(4x^2 - 3\right)^{\frac{3}{2}} + C$$

313. Make the following substitution:

$$u = 4x^5 - 10x^2 + 3$$

$$du = \left(20x^4 - 20x\right)dx \implies -\tfrac{1}{20}du = \left(x - x^4\right)dx$$

Applying this substitution in the integrand and computing the resulting indefinite integral yields

$$\int \frac{x - x^4}{4x^5 - 10x^2 + 3}\,dx = \int \frac{1}{4x^5 - 10x^2 + 3}\cdot\left(x - x^4\right)dx$$

$$= \int \tfrac{1}{u}\cdot\left(-\tfrac{1}{20}\right)du = -\tfrac{1}{20}\int \tfrac{1}{u}\,du = -\tfrac{1}{20}\ln|u| + C$$

Finally, rewrite the final expression of the preceding in terms of the original variable x by resubstituting $u = 4x^5 - 10x^2 + 3$ to obtain

$$\int \frac{x - x^4}{4x^5 - 10x^2 + 3}\,dx = -\tfrac{1}{20}\ln\left|4x^5 - 10x^2 + 3\right| + C$$

314. Make the following substitution:

$$u = \cos x$$

$$du = -\sin x\,dx \implies -du = \sin x\,dx$$

Applying this substitution in the integrand and computing the resulting indefinite integral yields

$$\int \cos^2 x\,\sin x\,dx = \int u^2(-1)du = -\int u^2\,du = -\tfrac{1}{3}u^3 + C$$

Finally, rewrite the final expression of the preceding equation in terms of the original variable x by resubstituting $u = \cos x$ to obtain

$$\int \cos^2 x\,\sin x\,dx = -\tfrac{1}{3}\cos^3 x + C$$

315. Make the following substitution:

$$u = 3x^3 + 1$$

$$du = 9x^2\,dx \implies \tfrac{1}{9}du = x^2\,dx$$

Applying this substitution in the integrand and computing the resulting indefinite integral yields

$$\int x^2 e^{3x^3+1}\,dx = \int e^{3x^3+1}x^2\,dx = \int e^u\left(\tfrac{1}{9}\right)du = \tfrac{1}{9}\int e^u\,du = \tfrac{1}{9}e^u + C$$

Finally, rewrite the final expression of the preceding equation in terms of the original variable x by resubstituting $u = 3x^3 + 1$ to obtain

$$\int x^2 e^{3x^3+1}\,dx = \tfrac{1}{9}e^{3x^3+1} + C$$

316. Make the following substitution:

$u = 2 - e^{3x}$

$du = -3e^{3x}\,dx \implies -\frac{1}{3}du = e^{3x}\,dx$

Applying this substitution in the integrand and computing the resulting indefinite integral yields

$\int \frac{e^{3x}}{2-e^{3x}}\,dx = \int \frac{1}{2-e^{3x}}\left(e^{3x}\right)dx = \int \frac{1}{u}\left(-\frac{1}{3}\right)du = -\frac{1}{3}\int \frac{1}{u}\,du = -\frac{1}{3}\ln|u| + C$

Finally, rewrite the final expression of the preceding equation in terms of the original variable x by resubstituting $u = 2 - e^{3x}$ to obtain

$\int \frac{e^{3x}}{2-e^{3x}}\,dx = -\frac{1}{3}\ln\left|2 - e^{3x}\right| + C$

317. Make the following substitution:

$u = \tan\left(\sqrt{x}+1\right)$

$du = \frac{\sec^2\left(\sqrt{x}+1\right)}{2\sqrt{x}}\,dx \implies 2\,du = \frac{\sec^2\left(\sqrt{x}+1\right)}{\sqrt{x}}\,dx$

Applying this substitution in the integrand and computing the resulting indefinite integral yields

$\int \frac{\tan\left(\sqrt{x}+1\right)\sec^2\left(\sqrt{x}+1\right)}{\sqrt{x}}\,dx = \int \tan\left(\sqrt{x}+1\right)\cdot \frac{\sec^2\left(\sqrt{x}+1\right)}{\sqrt{x}}\,dx$

$= \int u(2)\,du = 2\int u\,du = 2\cdot\frac{1}{2}u^2 + C = u^2 + C$

Finally, rewrite the final expression of the preceding equation in terms of the original variable x by resubstituting $u = \tan\left(\sqrt{x}+1\right)$ to obtain

$\int \frac{\tan\left(\sqrt{x}+1\right)\sec^2\left(\sqrt{x}+1\right)}{\sqrt{x}}\,dx = \tan^2\left(\sqrt{x}+1\right) + C$

318. First, rewrite the integral in the following equivalent manner:

$\int \frac{3x}{1+4x^4}\,dx = 3\int \frac{x}{1+\left(2x^2\right)^2}\,dx$

Make the following substitution:

$u = 2x^2$

$du = 4x\,dx \implies \frac{1}{4}du = x\,dx$

Applying this substitution in the integrand and computing the resulting indefinite integral yields

$\int \frac{3x}{1+4x^4}\,dx = 3\int \frac{x}{1+\left(2x^2\right)^2}\,dx = 3\int \frac{1}{1+u^2}\left(\frac{1}{4}\right)du = \frac{3}{4}\int \frac{1}{1+u^2}\,du = \frac{3}{4}\arctan u + C$

Finally, rewrite the final expression of the preceding equation in terms of the original variable x by resubstituting $u = 2x^2$ to obtain

$\int \frac{3x}{1+4x^4}\,dx = \frac{3}{4}\arctan\left(2x^2\right) + C$

319. Make the following substitution:

$u = e^x$

$du = e^x \, dx$

Applying this substitution in the integrand, using the trigonometric identity $1 - \sin^2 u = \cos^2 u$, and then computing the resulting indefinite integral yields

$$\int \frac{e^x}{1 - \sin^2\left(e^x\right)} \, dx = \int \frac{1}{1 - \sin^2\left(e^x\right)} \cdot e^x \, dx = \int \frac{1}{1 - \sin^2 u} \, du = \int \frac{1}{\cos^2 u} \, du$$

$$= \int \sec^2 u \, du = \tan u + C$$

Finally, rewrite the final expression of the preceding equation in terms of the original variable x by resubstituting $u = e^x$ to obtain

$$\int \frac{e^x}{1 - \sin^2\left(e^x\right)} \, dx = \tan\left(e^x\right) + C$$

320. First, rewrite the integral in the following equivalent manner:

$$\int \cot(\pi x) \, dx = \int \frac{\cos(\pi x)}{\sin(\pi x)} \, dx$$

Make the following substitution:

$u = \sin(\pi x)$

$du = \pi \cos(\pi x) \, dx \implies \frac{1}{\pi} du = \cos(\pi x) \, dx$

Applying this substitution in the integrand and computing the resulting indefinite integral yields

$$\int \cot(\pi x) \, dx = \int \frac{\cos(\pi x)}{\sin(\pi x)} \, dx = \int \frac{1}{\sin(\pi x)} \cdot \cos(\pi x) \, dx = \frac{1}{\pi} \int \frac{1}{u} \, du = \frac{1}{\pi} \ln|u| + C$$

Finally, rewrite the final expression of the preceding equation in terms of the original variable x by resubstituting $u = \sin(\pi x)$ to obtain

$$\int \cot(\pi x) \, dx = \frac{1}{\pi} \ln|\sin(\pi x)| + C$$

321. Make the following substitution:

$u = \ln(x+1)$

$du = \frac{1}{x+1} dx$

Applying this substitution in the integrand and computing the resulting indefinite integral yields

$$\int \frac{\sqrt[5]{\ln(x+1)}}{x+1} dx = \int \sqrt[5]{\ln(x+1)} \cdot \frac{1}{x+1} dx = \int \sqrt[5]{u}\, du = \int u^{\frac{1}{5}}\, du = \frac{u^{\frac{6}{5}}}{\frac{6}{5}} + C$$

$$= \frac{5}{6} u^{\frac{6}{5}} + C$$

Finally, rewrite the final expression of the preceding equation in terms of the original variable x by resubstituting $u = \ln(x + 1)$ to obtain

$$\int \frac{\sqrt[5]{\ln(x+1)}}{x+1} dx = \frac{5}{6}\left(\ln(x+1) \right)^{\frac{6}{5}} + C$$

322. First, rewrite the integral in the following equivalent manner:

$$\int \frac{2}{e^{\sqrt[3]{x}} \cdot \sqrt[3]{x^2}} dx = 2\int \frac{e^{-\sqrt[3]{x}}}{\sqrt[3]{x^2}} dx$$

Make the following substitution:

$u = -\sqrt[3]{x}$

$du = -\frac{1}{3\sqrt[3]{x^2}} dx \;\Rightarrow\; -3\,du = \frac{1}{\sqrt[3]{x^2}} dx$

Applying this substitution in the integrand and computing the resulting indefinite integral yields

$$2\int \frac{e^{-\sqrt[3]{x}}}{\sqrt[3]{x^2}} dx = 2\int e^{-\sqrt[3]{x}} \cdot \frac{1}{\sqrt[3]{x^2}} dx = 2\int e^u (-3)\,du = -6\int e^u\, du = -6e^u + C$$

Finally, rewrite the final expression of the preceding equation in terms of the original variable x by resubstituting $u = -\sqrt[3]{x}$ to obtain

$$\int \frac{2}{e^{\sqrt[3]{x}} \cdot \sqrt[3]{x^2}} dx = 6e^{-\sqrt[3]{x}} + C$$

323. First, apply the exponent rules to rewrite the integral in the following equivalent manner:

$$\int \frac{e^{3x+6}\left(e^x\right)^3}{e^{2x+1}}\,dx = \int \frac{e^{3x+6}e^{3x}}{e^{2x+1}}\,dx = \int \frac{e^{6x+6}}{e^{2x+1}}\,dx = \int e^{4x+5}\,dx$$

Make the following substitution:

$u = 4x + 5$

$$du = 4\,dx \;\Rightarrow\; \tfrac{1}{4}\,du = dx$$

Applying this substitution in the integrand and computing the resulting indefinite integral yields

$$\int \frac{e^{3x+6}\left(e^x\right)^3}{e^{2x+1}}\,dx = \int e^{4x+5}\,dx = \int e^u\left(\tfrac{1}{4}\right)du = \tfrac{1}{4}\int e^u\,du = \tfrac{1}{4}e^u + C$$

Finally, rewrite the final expression of the preceding equation in terms of the original variable x by resubstituting $u = 4x + 5$ to obtain

$$\int \frac{e^{3x+6}\left(e^x\right)^3}{e^{2x+1}}\,dx = \tfrac{1}{4}e^{4x+5} + C$$

324. Make the following substitution:

$$u = \frac{1}{x^3 + 8} = \left(x^3 + 8\right)^{-1}$$

$$du = -\left(x^3 + 8\right)^{-2}3x^2\,dx \;\Rightarrow\; -\tfrac{1}{3}\,du = \frac{x^2}{\left(x^3 + 8\right)^2}\,dx$$

Applying this substitution in the integrand and computing the resulting indefinite integral yields

$$\int \frac{x^2 \cos\left(\dfrac{1}{x^3+8}\right)}{\left(x^3+8\right)^2}\,dx = \int \cos\left(\frac{1}{x^3+8}\right)\cdot\frac{x^2}{\left(x^3+8\right)^2}\,dx = \int \cos u\left(-\tfrac{1}{3}\right)du$$

$$= -\tfrac{1}{3}\int \cos u\,du = -\tfrac{1}{3}\sin u + C$$

Finally, rewrite the final expression of the preceding equation in terms of the original variable x by resubstituting $u = \dfrac{1}{x^3 + 8}$ to obtain

$$\int \frac{x^2 \cos\left(\dfrac{1}{x^3+8}\right)}{\left(x^3+8\right)^2}\,dx = -\tfrac{1}{3}\sin\left(\frac{1}{x^3+8}\right) + C$$

325. Apply the formula for integration by parts $\int u\,dv = uv - \int v\,du$ with the following choices of u and v, along with their differentials:

$u = x \qquad dv = \cos(3x)dx$

$du = dx \qquad v = \int \cos(3x)dx = \frac{1}{3}\sin(3x)$

Applying the integration by parts formula yields:

$\int x\cos(3x)dx = (x)\left(\frac{1}{3}\sin(3x)\right) - \frac{1}{3}\int \sin(3x)dx$

$$= \frac{1}{3}x\sin(3x) - \frac{1}{3}\left(-\frac{1}{3}\cos(3x)\right) + C$$

$$= \frac{1}{3}x\sin(3x) + \frac{1}{9}\cos(3x) + C$$

Note: Computing both $\int \cos(3x)dx$ and $\int \sin(3x)dx$ entails using the substitution $z = 3x$, $\frac{1}{3}dz = dx$.

326. Apply the formula for integration by parts $\int u\,dv = uv - \int v\,du$ with the following choices of u and v, along with their differentials:

$u = x^2 + 2x + 5 \qquad dv = e^{-2x}dx$

$du = (2x+2)dx \qquad v = \int e^{-2x}\,dx = -\frac{1}{2}e^{-2x}$

Applying the integration by parts formula yields

$$\int (x^2 + 2x + 5)e^{-2x}\,dx = (x^2 + 2x + 5)\left(-\frac{1}{2}e^{-2x}\right) - \int \left(-\frac{1}{2}e^{-2x}\right)(2x+2)\,dx$$

$$= -\frac{1}{2}e^{-2x}(x^2 + 2x + 5) + \int e^{-2x}(x+1)\,dx$$

Computing the integral term in this equality requires another application of the integration by parts formula with the following new choices of u and v, along with their differentials:

$u = x + 1 \qquad dv = e^{-2x}dx$

$du = dx \qquad v = \int e^{-2x}\,dx = -\frac{1}{2}e^{-2x}$

Applying the parts formula yields

$$\int e^{-2x}(x+1)\,dx = (x+1)\left(-\frac{1}{2}e^{-2x}\right) - \int \left(-\frac{1}{2}e^{-2x}\right)dx$$

$$= -\frac{1}{2}e^{-2x}(x+1) + \frac{1}{2}\int e^{-2x}\,dx$$

$$= -\frac{1}{2}e^{-2x}(x+1) + \frac{1}{2}\left(-\frac{1}{2}e^{-2x}\right) + C$$

$$= -\frac{1}{2}e^{-2x}\left(x + 1 - \frac{1}{2}\right) + C$$

$$= -\frac{1}{2}e^{-2x}\left(x + \frac{1}{2}\right) + C$$

Substituting this back into the outcome of the first application of the integration by parts formula yields

$$\int (x^2 + 2x + 5)e^{-2x}\,dx = -\frac{1}{2}e^{-2x}(x^2 + 2x + 5) + \left(-\frac{1}{2}e^{-2x}\left(x + \frac{1}{2}\right) + C\right)$$

$$= -\frac{1}{2}e^{-2x}\left(x^2 + 2x + 5 + \left(x + \frac{1}{2}\right)\right) + C$$

$$= -\frac{1}{2}e^{-2x}\left(x^2 + 3x + \frac{11}{2}\right) + C$$

327. We first apply the substitution technique to simplify the integral.
Precisely, make the following substitution:

$z = \sin(2x)$

$dz = 2\cos(2x)dx \implies \frac{1}{2}dz = \cos(2x)dx$

Applying this substitution yields the following equivalent integral:

$\int \cos(2x) \cdot \ln(\sin(2x))dx = \frac{1}{2}\int \ln z \, dz$

Now, apply the formula for integration by parts $\int u \, dv = uv - \int v \, du$ with the

following choices of u and v, along with their differentials:

$u = \ln z \qquad dv = dz$

$du = \frac{1}{z}dz \qquad v = \int dz = z$

Applying the integration by parts formula yields:

$\int \ln z \, dz = z \ln z - \int z\left(\frac{1}{z}\right)dz = z \ln z - \int dz = z \ln z - z + C = z(\ln z - 1) + C$

Substituting this back into the equality obtained from our initial step of
applying the substitution technique, and subsequently resubstituting
$z = \sin(2x)$, yields

$\int \cos(2x) \cdot \ln(\sin(2x))dx = \frac{1}{2}(z \ln z - z) + C = \frac{1}{2}z(\ln z - 1) + C$

$= \frac{1}{2}\sin(2x)\left(\ln(\sin(2x)) - 1\right) + C$

328. This problem will require two successive applications of integration by parts, followed by a clever algebraic manipulation. To begin, apply the formula for integration by parts $\int u\,dv = uv - \int v\,du$ with the following choices of u and v, along with their differentials:

$$u = e^{-x} \qquad dv = \sin x\,dx$$

$$du = -e^{-x}dx \qquad v = \int \sin x\,dx = -\cos x$$

Applying the integration by parts formula yields:

$$\int e^{-x}\sin x\,dx = \left(e^{-x}\right)(-\cos x) - \int(-\cos x)\left(-e^{-x}\right)dx$$

$$= -e^{-x}\cos x - \int e^{-x}\cos x\,dx$$

Observe that the integral on the right side of this equality is similar in nature to the original integral. We apply the integration by parts formula with the following new choices of u and v, along with their differentials, to compute $\int e^{-x}\cos x\,dx$:

$$u = e^{-x} \qquad dv = \cos x\,dx$$

$$du = -e^{-x}dx \qquad v = \int \cos x\,dx = \sin x$$

Applying the integration by parts formula yields:

$$\int e^{-x}\cos x\,dx = \left(e^{-x}\right)(\sin x) - \int(\sin x)\left(-e^{-x}\right)dx$$

$$= e^{-x}\sin x + \int e^{-x}\sin x\,dx + C$$

Note that the original integral now appears on the right side of the equality. While this at first seems circular, substitute the expression for $\int e^{-x}\cos x\,dx$ into the equality resulting from the first application of integration by parts to obtain:

$$\int e^{-x}\sin x\,dx = -e^{-x}\cos x - \int e^{-x}\cos x\,dx$$

$$= -e^{-x}\cos x - \left(e^{-x}\sin x + \int e^{-x}\sin x\,dx + C\right)$$

$$= \left(-e^{-x}\cos x - e^{-x}\sin x\right) - \int e^{-x}\sin x\,dx + C$$

Here's where the clever algebraic trick comes into play. Move the integral term that is at present on the right side of the equality to the left, and combine with the one already there to obtain the following equivalent equality:

$$2\int e^{-x}\sin x\,dx = \left(-e^{-x}\cos x - e^{-x}\sin x\right) + C$$

Now, simply divide both sides by 2 to obtain the following expression, devoid of additional integrals, that is equivalent to the original integral:

$$\int e^{-x}\sin x\,dx = \tfrac{1}{2}\left(-e^{-x}\cos x - e^{-x}\sin x\right)+C = -\tfrac{e^{-x}}{2}(\cos x + \sin x)+C$$

(Note: An equally valid approach would be to reverse the identifications of u and v in both applications of the integration by parts formula. Doing so results in the same antiderivative.)

329. We first apply the substitution technique to simplify the integral. Make the following substitution:

$$z = \ln\left(e^x + 1\right)$$

$$dz = \frac{e^x}{e^x + 1}\,dx$$

Applying this substitution yields the following equivalent integral:

$$\int \frac{3\ln\left(\ln\left(e^x + 1\right)\right)e^x}{e^x + 1}\,dx = 3\int \ln z\,dz$$

Now, apply the formula for integration by parts $\int u\,dv = uv - \int v\,du$ with the following choices of u and v, along with their differentials:

$$u = \ln z \qquad dv = dz$$

$$du = \tfrac{1}{z}\,dz \qquad v = \int dz = z$$

Applying the integration by parts formula yields:

$$\int \ln z\,dz = z\ln z - \int z\left(\tfrac{1}{z}\right)dz = z\ln z - \int dz = z\ln z - z + C = z(\ln z - 1) + C$$

Substituting this back into the equality obtained from our initial step of applying the substitution technique, and subsequently resubstituting $z = \ln\left(e^x + 1\right)$, yields

$$\int \frac{3\ln\left(\ln\left(e^x + 1\right)\right)e^x}{e^x + 1}\,dx = 3\int \ln z\,dz = 3z(\ln z - 1) + C$$

$$= 3\ln\left(e^x + 1\right)\cdot\left(\ln\left(\ln\left(e^x + 1\right)\right) - 1\right) + C$$

330. Apply the formula for integration by parts $\int u\,dv = uv - \int v\,du$ with the following choices of u and v, along with their differentials:

$$u = \arctan(4x) \qquad dv = dx$$

$$du = \frac{4}{1+(4x)^2}\,dx \qquad v = \int dx = x$$

Applying the integration by parts formula yields:

$$\int \arctan(4x)\,dx = x\arctan(4x) - \int \frac{4x}{1+(4x)^2}\,dx = x\arctan(4x) - \int \frac{4x}{1+16x^2}\,dx$$

Now, to compute the integral on the right side of the equality, make the following substitution:

$$z = 1 + 16x^2$$

$$dz = 32x\,dx \;\Rightarrow\; \tfrac{1}{8}dz = 4x\,dx$$

Applying this substitution in the integrand and computing the resulting indefinite integral yields

$$\int \frac{4x}{1+16x^2}\,dx = \int \frac{\frac{1}{8}}{z}\,dz = \tfrac{1}{8}\int \tfrac{1}{z}\,dz = \tfrac{1}{8}\ln|z| + C$$

Substituting this back into the equality obtained from our initial step of applying the substitution technique, and subsequently resubstituting $z = 1 + 16x^2$, yields

$$\int \arctan(4x)\,dx = x\arctan(4x) - \int \frac{4x}{1+16x^2}\,dx$$

$$= x\arctan(4x) - \tfrac{1}{8}\ln|z| + C$$

$$= x\arctan(4x) - \tfrac{1}{8}\ln\left|1+16x^2\right| + C$$

$$= x\arctan(4x) - \tfrac{1}{8}\ln\left(1+16x^2\right) + C$$

331. First, apply the method of partial fraction decomposition to rewrite the integrand in a more readily integrable form. The partial fraction decomposition has the form:

$$\frac{1}{x(x-1)} = \frac{A}{x} + \frac{B}{x-1}$$

To find the coefficients, multiply both sides of the equality by $x(x-1)$ and gather like terms to obtain

$$1 = A(x-1) + Bx$$

$$1 = (A+B)x - A$$

Now, equate corresponding coefficients in the equality to obtain the following system of equations whose unknowns are the coefficients we seek:

$$\begin{cases} A+B=0 \\ \quad -A=1 \end{cases}$$

Now, solve this system. Substituting $A = -1$ into the first equation of the system yields $B = 1$. Thus, the partial fraction decomposition becomes

$$\frac{1}{x(x-1)} = \frac{1}{x-1} - \frac{1}{x}$$

We substitute this expression in for the integrand in the original integral, compute each, and simplify, as follows:

$$\int \frac{1}{x(x-1)} dx = \int \left(\frac{1}{x-1} - \frac{1}{x}\right) dx = \int \frac{1}{x-1} dx - \int \frac{1}{x} dx$$

$$= \ln|x-1| - \ln|x| + C = \ln\left|\frac{x-1}{x}\right| + C$$

(Note: A substitution of $u = x - 1$ is used to compute $\int \frac{1}{x-1} dx$. Also, the logarithm of a quotient rule is used to obtain the simplified expression on the right side of the preceding expression.)

332. First, apply the method of partial fraction decomposition to rewrite the integrand in a more readily integrable form. The partial fraction decomposition has the form:

$$\frac{9x-2}{(x-1)^2} = \frac{A}{x-1} + \frac{B}{(x-1)^2}$$

To find the coefficients, multiply both sides of the equality by $(x-1)^2$ and gather like terms to obtain

$$9x-2 = A(x-1) + B$$

$$9x-2 = Ax + (B-A)$$

Now, equate corresponding coefficients in the equality to obtain the following system of equations whose unknowns are the coefficients we seek:

$$\begin{cases} A = 9 \\ B - A = -2 \end{cases}$$

Now, solve this system. Substituting $A = 9$ into the second equation of the system yields $B = 7$. Thus, the partial fraction decomposition becomes

$$\frac{9x-2}{(x-1)^2} = \frac{9}{x-1} + \frac{7}{(x-1)^2}$$

We substitute this expression in for the integrand in the original integral, compute each, and simplify, as follows:

$$\int \frac{9x-2}{(x-1)^2}\,dx = \int \left(\frac{9}{x-1} + \frac{7}{(x-1)^2} \right)dx = 9\int \frac{1}{x-1}\,dx + 7\int (x-1)^{-2}\,dx$$

$$= 9\ln|x-1| - 7(x-1)^{-1} + C = 9\ln|x-1| - \frac{7}{x-1} + C$$

333. First, apply the method of partial fraction decomposition to rewrite the integrand in a more readily integrable form. The partial fraction decomposition has the form:

$$\frac{9x-11}{(x-3)(x+5)} = \frac{A}{x-3} + \frac{B}{x+5}$$

To find the coefficients, multiply both sides of the equality by $(x-3)(x+5)$ and gather like terms to obtain

$$9x - 11 = A(x+5) + B(x-3)$$

$$9x - 11 = (A+B)x + (5A-3B)$$

Now, equate corresponding coefficients in the equality to obtain the following system of equations whose unknowns are the coefficients we seek:

$$\begin{cases} A+B=9 \\ 5A-3B=-11 \end{cases}$$

Now, solve this system. Multiply the first equation by –5 to obtain $-5A - 5B = -45$; add this to the second equation and solve for B to obtain

$$-8B = -56 \implies B = 7$$

Substituting this into the first equation yields $A = 2$. Thus, the partial fraction decomposition becomes

$$\frac{9x-11}{(x-3)(x+5)} = \frac{2}{x-3} + \frac{7}{x+5}$$

We substitute this expression in for the integrand in the original integral, compute each, and simplify, as follows:

$$\int \frac{9x-11}{(x-3)(x+5)}dx = \int \left(\frac{2}{x-3} + \frac{7}{x+5}\right)dx = 2\int \frac{1}{x-3}dx + 7\int \frac{1}{x+5}dx$$

$$= 2\ln|x-3| + 7\ln|x+5| + C$$

Applying the logarithm rules enables us to simplify this expression as follows:

$$2\ln|x-3| + 7\ln|x+5| = \ln|x-3|^2 + \ln|x+5|^7 = \ln\left(|x-3|^2|x+5|^7\right)$$

Thus, we conclude that

$$\int \frac{9x-11}{(x-3)(x+5)}dx = \ln\left(|x-3|^2|x+5|^7\right) + C$$

334. First, apply the method of partial fraction decomposition to rewrite the integrand in a more readily integrable form. The partial fraction decomposition has the form:

$$\frac{x^3}{\left(x^2+9\right)^2} = \frac{Ax+B}{x^2+9} + \frac{Cx+D}{\left(x^2+9\right)^2}$$

To find the coefficients, multiply both sides of the equality by $(x^2+9)^2$ and gather like terms to obtain

$$x^3 = \left(Ax+B\right)\left(x^2+9\right)+\left(Cx+D\right) = Ax^3 + Bx^2 + 9Ax + 9B + Cx + D$$
$$= Ax^3 + Bx^2 + \left(9A+C\right)x + \left(9B+D\right)$$

Now, equate corresponding coefficients in the equality to obtain the following system of equations whose unknowns are the coefficients we seek:

$$\begin{cases} A=1 \\ B=0 \\ 9A+C=0 \\ 9B+D=0 \end{cases}$$

Now, solve this system. Substitute $A=1$ into the third equation to obtain $C=-9$, and substitute $B=0$ into the fourth equation to conclude that $D=0$. As such, the partial fraction decomposition becomes:

$$\frac{x^3}{\left(x^2+9\right)^2} = \frac{x}{x^2+9} - \frac{9x}{\left(x^2+9\right)^2}$$

We substitute this expression in for the integrand in the original integral to obtain

$$\int \frac{x^3}{\left(x^2+9\right)^2}dx = \int \left(\frac{x}{x^2+9} - \frac{9x}{\left(x^2+9\right)^2}\right)dx = \int \frac{x}{x^2+9}dx - 9\int \frac{x}{\left(x^2+9\right)^2}dx$$

Now, compute each of the two integrals on the right side using appropriate substitutions. For both integrals, use the following substitution:

$$u = x^2+9$$
$$du = 2x\,dx \implies \tfrac{1}{2}du = x\,dx$$

Applying this substitution in each integral and computing the resulting indefinite integrals yields

$$\int \frac{x}{x^2+9}\,dx = \int \frac{1}{u}\left(\frac{1}{2}\right)du = \frac{1}{2}\int \frac{1}{u}\,du = \frac{1}{2}\ln|u| + C = \frac{1}{2}\ln\left|x^2+9\right| + C$$

$$\int \frac{x}{\left(x^2+9\right)^2}\,dx = \int \frac{1}{u^2}\left(\frac{1}{2}\right)du = \frac{1}{2}\int \frac{1}{u^2}\,du = \frac{1}{2}\int u^{-2}\,du = -\frac{1}{2}u^{-1} + C = -\frac{1}{2u} + C$$

$$= -\frac{1}{2\left(x^2+9\right)} + C$$

Hence, we conclude that

$$\int \frac{x^3}{\left(x^2+9\right)^2}\,dx = \int \frac{x}{x^2+9}\,dx - 9\int \frac{x}{\left(x^2+9\right)^2}\,dx$$

$$= \frac{1}{2}\ln\left|x^2+9\right| - 9\left(-\frac{1}{2\left(x^2+9\right)}\right) + C$$

$$= \frac{1}{2}\ln\left|x^2+9\right| + \frac{9}{2\left(x^2+9\right)} + C$$

$$= \frac{1}{2}\ln\left(x^2+9\right) + \frac{9}{2\left(x^2+9\right)} + C$$

335. The trick is to rewrite the integrand in a manner in which the trigonometric identity $\sin^2\theta + \cos^2\theta = 1$ (equivalently, $\sin^2\theta = 1 - \cos^2\theta$) can be used advantageously, as follows:

$$\int \sin^3\left(\tfrac{\pi}{2}x\right)dx = \int \sin\left(\tfrac{\pi}{2}x\right)\sin^2\left(\tfrac{\pi}{2}x\right)dx = \int \sin\left(\tfrac{\pi}{2}x\right)\left[1 - \cos^2\left(\tfrac{\pi}{2}x\right)\right]dx$$

$$= \int \sin\left(\tfrac{\pi}{2}x\right)dx - \int \sin\left(\tfrac{\pi}{2}x\right)\cos^2\left(\tfrac{\pi}{2}x\right)dx$$

Now, we apply the substitution method to compute each of the integrals on the right side of the equality. For the first integral, use the following substitution:

$$u = \tfrac{\pi}{2}x$$

$$du = \tfrac{\pi}{2}dx \implies \tfrac{2}{\pi}du = dx$$

Applying this substitution and computing the resulting indefinite integral yields

$$\int \sin\left(\tfrac{\pi}{2}x\right)dx = \int \sin u\left(\tfrac{2}{\pi}\right)du = \tfrac{2}{\pi}\int \sin u\, du = \tfrac{2}{\pi}(-\cos u) + C$$

$$= -\tfrac{2}{\pi}\cos\left(\tfrac{\pi}{2}x\right) + C$$

For the second integral, use the following substitution:

$$u = \cos\left(\tfrac{\pi}{2}x\right)$$

$$du = -\tfrac{\pi}{2}\sin\left(\tfrac{\pi}{2}x\right)dx \implies -\tfrac{2}{\pi}du = \sin\left(\tfrac{\pi}{2}x\right)dx$$

Applying this substitution and computing the resulting indefinite integral yields

$$\int \sin\left(\tfrac{\pi}{2}x\right)\cos^2\left(\tfrac{\pi}{2}x\right)dx = \int u^2\left(\tfrac{-2}{\pi}\right)du = -\tfrac{2}{\pi}\int u^2\, du$$

$$= -\tfrac{2}{\pi}\cdot\tfrac{u^3}{3} + C = -\tfrac{2}{3\pi}\cos^3\left(\tfrac{\pi}{2}x\right) + C$$

Hence, we conclude that

$$\int \sin^3\left(\tfrac{\pi}{2}x\right)dx = \int \sin\left(\tfrac{\pi}{2}x\right)dx - \int \sin\left(\tfrac{\pi}{2}x\right)\cos^2\left(\tfrac{\pi}{2}x\right)dx$$

$$= -\tfrac{2}{\pi}\cos\left(\tfrac{\pi}{2}x\right) - \left(-\tfrac{2}{3\pi}\cos^3\left(\tfrac{\pi}{2}x\right)\right) + C$$

$$= -\tfrac{2}{\pi}\cos\left(\tfrac{\pi}{2}x\right) + \tfrac{2}{3\pi}\cos^3\left(\tfrac{\pi}{2}x\right) + C$$

$$= -\tfrac{2}{\pi}\cos\left(\tfrac{\pi}{2}x\right)\left(1 - \tfrac{1}{3}\cos^2\left(\tfrac{\pi}{2}x\right)\right) + C$$

336. The trick is to apply the double-angle formula $\sin^2 \theta = \frac{1 - \cos 2\theta}{2}$ with

$\theta = 3x$, and compute the indefinite integrals that follow:

$$\int \sin^2(3x)\,dx = \int \frac{1 - \cos(6x)}{2}\,dx = \frac{1}{2}\left[\int 1\,dx - \int \cos(6x)\,dx\right]$$

$$= \frac{1}{2}\left[x - \frac{1}{6}\sin(6x)\right] + C$$

337. The trick is to rewrite the integrand in a manner in which the trigonometric identity $\sin^2 \theta + \cos^2 \theta = 1$ (equivalently, $\sin^2 \theta = 1 - \cos^2 \theta$) can be used advantageously, as follows:

$$\int \sin^2 x \cos^3 x\,dx = \int \sin^2 x \cos^2 x \cos x\,dx = \int \sin^2 x\left(1 - \sin^2 x\right)\cos x\,dx$$

$$= \int \left(\sin^2 x - \sin^4 x\right)\cos x\,dx$$

Now, make the following substitution:

$u = \sin x$

$du = \cos x\,dx$

Applying this substitution and computing the resulting indefinite integral yields

$$\int \sin^2 x \cos^3 x\,dx = \int \left(\sin^2 x - \sin^4 x\right)\cos x\,dx = \int \left(u^2 - u^4\right)du$$

$$= \frac{1}{3}u^3 - \frac{1}{5}u^5 + C = \frac{1}{3}\sin^3 x - \frac{1}{5}\sin^5 x + C$$

338. The trick is to rewrite the integrand in such a manner that, upon making a suitable u-substitution, the integrand becomes $\frac{1}{1+u^2}$, the antiderivative of which is arctan$u$. To this end, we proceed as follows:

$$\int \frac{2}{3+7x^2}\,dx = \int \frac{2}{3\left(1+\frac{7}{3}x^2\right)}\,dx = \frac{2}{3}\int \frac{1}{1+\left(\sqrt{\frac{7}{3}}\,x\right)^2}\,dx$$

Now, make the following substitution:

$$u = \sqrt{\frac{7}{3}}\,x$$

$$du = \sqrt{\frac{7}{3}}\,dx \;\Rightarrow\; \sqrt{\frac{3}{7}}\,du = dx$$

Applying this substitution and computing the resulting indefinite integral yields

$$\int \frac{2}{3+7x^2}\,dx = \frac{2}{3}\int \frac{1}{1+\left(\sqrt{\frac{7}{3}}\,x\right)^2}\,dx = \frac{2}{3}\int \frac{\sqrt{\frac{3}{7}}}{1+u^2}\,du = \frac{2}{3}\sqrt{\frac{3}{7}}\int \frac{1}{1+u^2}\,du$$

$$= \frac{2}{\sqrt{21}}\arctan u + C$$

Resubstituting $u = \sqrt{\frac{7}{3}}\,x$ in this expression yields

$$\int \frac{2}{3+7x^2}\,dx = \frac{2}{\sqrt{21}}\arctan\left(\sqrt{\frac{7}{3}}\,x\right) + C$$

$$= \frac{2\sqrt{21}}{21}\arctan\left(\frac{\sqrt{21}}{3}\,x\right) + C$$

339. Simplify the integrand, as follows:

$$\frac{\left(x^3-1\right)^2\left(x^3+1\right)^2}{x^6}=\frac{\left[\left(x^3-1\right)\left(x^3+1\right)\right]^2}{x^6}=\frac{\left[x^6-1\right]^2}{x^6}=\frac{x^{12}-2x^6+1}{x^6}=x^6-2+x^{-6}$$

Now, substitute this equivalent expression in for the integrand in the original integral and compute the resulting indefinite integral to obtain

$$\int\frac{\left(x^3-1\right)^2\left(x^3+1\right)^2}{x^6}dx=\int\left(x^6-2+x^{-6}\right)dx$$
$$=\frac{x^7}{7}-2x+\frac{x^{-5}}{-5}+C=\frac{x^7}{7}-2x-\frac{1}{5x^5}+C$$

340. Make the following substitution:

$$u=x\ln x$$
$$du=\left[x\left(\tfrac{1}{x}\right)+(\ln x)(1)\right]dx=(1+\ln x)dx$$

Applying this substitution and computing the resulting indefinite integral yields

$$\int\frac{1+\ln x}{\sin^2(x\ln x)}dx=\int\frac{1}{\sin^2 u}du=\int\csc^2 u\,du=-\cot u+C$$

Resubstituting $u=x\ln x$ into this expression yields

$$\int\frac{1+\ln x}{\sin^2(x\ln x)}dx=-\cot(x\ln x)+C$$

14

Applications of Integration— Area Problems

The Riemann integral $\int_a^b f(x)\,dx$ is linked geometrically to the planar region formed using $y = f(x)$, $x = a$, $x = b$, and the x-axis. This number, in general, can be negative and therefore, does not truly represent the area of the region. However, slight adjustments can be made to obtain an integral that *does* represent the area of the region. There are three situations:

1. If the graph of $y = f(x)$ lies entirely above (or on) the x-axis for all x in the interval $[a,b]$, then the area of the region described in the preceding paragraph is indeed given by $\int_a^b f(x)\,dx$.

2. If the graph of $y = f(x)$ lies entirely below (or on) the x-axis for all x in the interval $[a,b]$, then the area of the region described above is given by $\int_a^b f(x)\,dx$.

3. If the graph of $y = f(x)$ is above the x-axis for all x in the interval $[a,c]$ and is below the x-axis for all x in the interval $[c,b]$, then the area of the region described above is given by $\int_a^b f(x)\,dx = \int_a^c f(x)\,dx - \int_c^b f(x)\,dx$.

These principles can be extended to regions whose boundaries no longer include the x-axis, but rather whose lower boundary is another function $y = g(x)$. In such case, if the graph of $y = f(x)$ lies above the graph of $y = g(x)$ for all x in the interval $[a,b]$, then the area of the region bounded by $y = f(x)$, $y = g(x)$, $x = a$, and $x = b$ is given by $\int_a^b \left[f(x) - g(x) \right] dx$. In words, you integrate the top curve minus the bottom curve over the interval formed using the leftmost x-value to the rightmost x-value used to form the region of integration. More elaborate versions of this can be formed if the top and/or bottom boundary curves are defined piecewise (meaning that the definitions of the curves change at some x-value in the interval $[a,b]$). In such case, the same principle applies, together with an application of interval additivity.

Finally, the rectangles formed in defining the integral leading to the area of a region can be either vertical (as is the case described earlier when the variable of integration is x and the interval of integration is taken on the x-axis) or horizontal. In case of the latter, the variable of integration is y, the region of integration is formed using curves solved for x in terms of y, say $x = F(y)$ and $x = G(y)$, and horizontal lines $y = c$ and $y = d$ (which give rise to the interval of integration). In words, you integrate the right curve minus the left curve over the interval formed using the bottommost y-value to the topmost y-value used to form the region of integration. As before, more elaborate situations occur when the boundary curves are piecewise-defined. The same approach is used, adapted to this "horizontal rectangle" setting.

Questions

341. The area of the following shaded region can be computed using which of the following integrals?

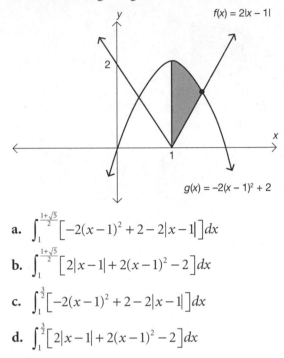

$f(x) = 2|x - 1|$

$g(x) = -2(x - 1)^2 + 2$

a. $\displaystyle\int_{1}^{\frac{1+\sqrt{5}}{2}}\left[-2(x-1)^2 + 2 - 2|x-1|\right]dx$

b. $\displaystyle\int_{1}^{\frac{1+\sqrt{5}}{2}}\left[2|x-1| + 2(x-1)^2 - 2\right]dx$

c. $\displaystyle\int_{1}^{\frac{3}{2}}\left[-2(x-1)^2 + 2 - 2|x-1|\right]dx$

d. $\displaystyle\int_{1}^{\frac{3}{2}}\left[2|x-1| + 2(x-1)^2 - 2\right]dx$

342. The area of the following shaded region can be computed using which of the following integrals?

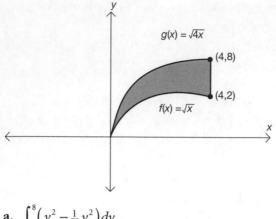

a. $\int_0^8 \left(y^2 - \frac{1}{4} y^2 \right) dy$

b. $\int_0^8 \left(4\sqrt{x} - \sqrt{x} \right) dx$

c. $\int_0^2 \left(y^2 - \frac{1}{4} y^2 \right) dy + \int_2^8 \left(4 - \frac{1}{4} y^2 \right) dy$

d. $\int_0^4 \left(\sqrt{x} - 4\sqrt{x} \right) dx$

343. The area of the following shaded region can be computed using which of the following integrals?

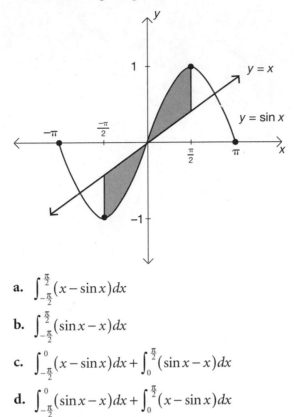

a. $\displaystyle\int_{-\frac{\pi}{2}}^{\frac{\pi}{2}}(x-\sin x)\,dx$

b. $\displaystyle\int_{-\frac{\pi}{2}}^{\frac{\pi}{2}}(\sin x-x)\,dx$

c. $\displaystyle\int_{-\frac{\pi}{2}}^{0}(x-\sin x)\,dx+\int_{0}^{\frac{\pi}{2}}(\sin x-x)\,dx$

d. $\displaystyle\int_{-\frac{\pi}{2}}^{0}(\sin x-x)\,dx+\int_{0}^{\frac{\pi}{2}}(x-\sin x)\,dx$

344. The area of the following shaded region can be computed using which of the following integrals?

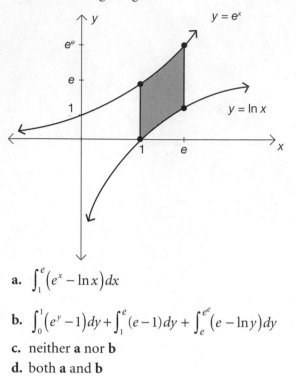

a. $\displaystyle\int_{1}^{e}\left(e^{x}-\ln x\right)dx$

b. $\displaystyle\int_{0}^{1}\left(e^{y}-1\right)dy+\int_{1}^{e}\left(e-1\right)dy+\int_{e}^{e^{e}}\left(e-\ln y\right)dy$

c. neither **a** nor **b**

d. both **a** and **b**

345. The area of the following shaded region can be computed using which of the following integrals?

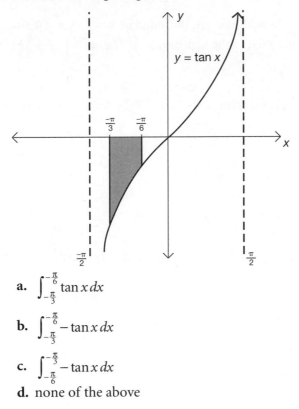

a. $\int_{-\frac{\pi}{3}}^{-\frac{\pi}{6}} \tan x\, dx$

b. $\int_{-\frac{\pi}{3}}^{-\frac{\pi}{6}} -\tan x\, dx$

c. $\int_{-\frac{\pi}{6}}^{-\frac{\pi}{3}} -\tan x\, dx$

d. none of the above

For Questions 346 through 350, use the following diagram:

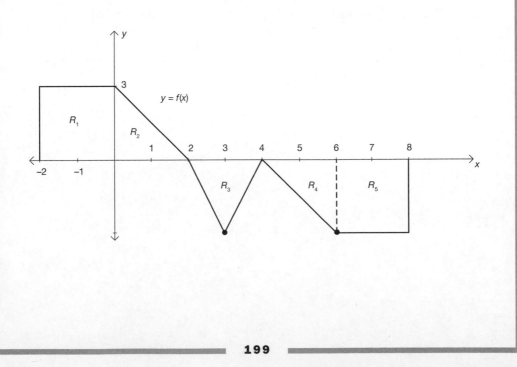

346. True or false? $\int_{-2}^{0} f(x)\,dx = \int_{6}^{8} f(x)\,dx$

347. True or false? The area of the region bounded by $y = f(x)$, $x = 0$ and $x = 6$ can be computed using the expression $\int_{0}^{2} f(x)\,dx - \int_{2}^{4} f(x)\,dx$ $-\int_{4}^{6} f(x)\,dx$.

348. True or false? $\int_{-2}^{8} f(x)\,dx = \text{Area}\left(R_1\right) + \text{Area}\left(R_2\right) + \text{Area}\left(R_3\right) + \text{Area}\left(R_4\right)$ $+ \text{Area}\left(R_5\right)$

349. True or false? $\int_{4}^{6} f(x)\,dx = \text{Area}\left(R_2\right)$

350. True or false? $\int_{-2}^{2} f(x)\,dx = -\int_{4}^{8} f(x)\,dx$

For Questions 351 through 355, determine the area of the described region.

351. The region bounded by $y = \cos x$, $x = \frac{\pi}{2}$, $x = \pi$, and the x-axis.

352. The region bounded by $y = e^{2x}$, $x = -\ln 3$, $x = \ln 2$, and the x-axis.

353. The region between $y = -x^3$ and $y = x^2$ in quadrant II.

354. The region bounded by $y = \ln(ex)$, $y = \ln x$, $x = 1$, and $x = e$.

355. The region bounded by $y = 3\sin x$, $y = \sin x$, $x = 0$, and $x = 2\pi$.

Answers

341. a. First, determine the x-value of the point of intersection of the graphs of f and g that occurs to the right of $x = 1$. For such values of x, we can replace $|x - 1|$ by simply $(x - 1)$. Taking this into account, equating the expressions for f and g and solving for x yields:

$$2(x-1) = -2(x-1)^2 + 2$$

$$2x - 2 = -2(x^2 - 2x + 1) + 2$$

$$2x - 2 = -2x^2 + 4x$$

$$2(x^2 - x - 1) = 0$$

$$x = \frac{1 \pm \sqrt{1+4}}{2} = \frac{1 \pm \sqrt{5}}{2}$$

The x-value of the rightmost intersection point is therefore $\frac{1+\sqrt{5}}{2}$.

Next, looking back at the region, we see that if we use vertical rectangles to generate the integral, the interval of integration should be $\left[1, \frac{1+\sqrt{5}}{2}\right]$ and the integrand is obtained by subtracting the top curve minus the bottom curve, which yields $-2(x-1)^2 + 2 - 2|x-1|$. Hence, the area of the region is given by $\int_1^{\frac{1+\sqrt{5}}{2}} \left[-2(x-1)^2 + 2 - 2|x-1|\right] dx$.

342. **c.** Using vertical rectangles would be the easier of the two approaches
to use to generate an integral expression used to compute the area of
the region. Looking back at the region, the interval of integration along
the x-axis would be $[0,4]$ and the integrand would be obtained by
subtracting the top curve minus the bottom curve, which yields
$4\sqrt{x} - \sqrt{x}$. Hence, the area of the region using vertical rectangles would
be $\int_0^4 (4\sqrt{x} - \sqrt{x})dx$. This is not a choice listed, however, so we must
determine an equivalent integral expression using horizontal rectangles.

We must solve the expressions used to define the functions for x
in terms of y. Doing so, we see that the boundaries of the region are
$x = y^2$, $x = \frac{1}{4}y^2$, and $x = 4$. The bottommost y-value used to generate
the region is $y = 0$ and the topmost is $y = 8$. However, when determining the heights of the horizontal rectangles by subtracting the rightmost
curve minus the leftmost curve, we see that the curve that plays the role
of the rightmost curve changes at $y = 2$. We must therefore use interval
additivity to split the region into two disjoint pieces, determine an
integral used to compute the areas of each of the two pieces separately,
and then add them together. Doing so yields
$\int_0^2 \left(y^2 - \frac{1}{4}y^2\right)dy + \int_2^8 \left(4 - \frac{1}{4}y^2\right)dy$, where the first integral corresponds to
the area of the lower of the two regions (i.e., below where the first
integral corresponds to the area of the lower of the two regions; i.e.,
below $y = 2$) and the second integral corresponds to the area of the
upper of the two regions.

343. c. Using vertical rectangles would be the easier of the two approaches to use to generate an integral expression used to compute the area of the region. Looking back at the region, the interval of integration along the x-axis would be $\left[-\frac{\pi}{2}, \frac{\pi}{2}\right]$. The integrand is obtained by subtracting the top curve minus the bottom curve; however, note that these curves change at $x = 0$. We therefore split the region into two disjoint pieces at $x = 0$, determine an integral for the areas of each of the two pieces separately, and then add them together. Doing so yields

$$\int_{-\frac{\pi}{2}}^{0}(x - \sin x)\,dx + \int_{0}^{\frac{\pi}{2}}(\sin x - x)\,dx$$

344. d. We begin by generating an integral expression *using vertical rectangles* that can be used to compute the area of the shaded region. The interval of integration (taken along the x-axis) is $[1,e]$, and the boundary curves of the region do not change throughout this interval. Thus, the area of the region using vertical rectangles is given by the single integral $\int_{1}^{e}(e^x - \ln x)\,dx$.

Next, we generate an integral expression *using horizontal rectangles* that can be used to compute the area of the shaded region. The interval of integration (taken along the y-axis) is $[0,e^e]$. This time, however, the curves identified as the rightmost and leftmost boundaries change at *two* places in this region, namely at $y = 1$ and $y = e$. (This can be easily seen by drawing horizontal lines through the region at these y-values.) In order to obtain the integrands, the expressions for the curves must be solved for x in terms of y: $x = \ln y$, $x = e^y$, $x = 1$, $x = e$. Now, we generate an integral for each of the three disjoint regions obtained by dividing the original region at $y = 1$ and $y = e$. Doing so yields the integral expression $\int_{0}^{1}(e^y - 1)\,dy + \int_{1}^{e}(e-1)\,dy + \int_{e}^{e^e}(e - \ln y)\,dy$. Thus, the integral expressions given by choices **a** and **b** can both be used to compute the area of the shaded region.

345. b. Using vertical rectangles is the easier of the two approaches to use to generate an integral expression used to compute the area of the region. Looking back at the region, the interval of integration along the x-axis is $\left[-\frac{\pi}{3}, -\frac{\pi}{6}\right]$. The integrand is obtained by subtracting the top curve minus the bottom curve, which yields $0 - \tan x$. Thus, the integral used to compute the area of this region is $\int_{-\frac{\pi}{3}}^{-\frac{\pi}{6}} - \tan x\,dx$.

346. False. The *areas* of regions R_1 and R_5 are the same, but the integrals are related by the equation $\int_{-2}^{0}f(x)\,dx = -\int_{6}^{8}f(x)\,dx$.

347. True. In general, if the graph of $y = f(x)$ is above the x-axis for all x in the interval $[a,c]$ and is below the x-axis for all x in the interval $[c,b]$, then the area of the region described is given by
$\int_a^b f(x)\,dx = \int_a^c f(x)\,dx - \int_c^b f(x)\,dx$. This principle applies for the given scenario with $a = 0$, $b = 6$, and $c = 2$. Indeed, using interval additivity, we can rewrite the given expression as follows:

$$\int_0^2 f(x)\,dx - \int_2^4 f(x)\,dx - \int_4^6 f(x)\,dx = \int_0^2 f(x)\,dx - \left[\int_2^4 f(x)\,dx + \int_4^6 f(x)\,dx\right]$$

$$= \int_0^2 f(x)\,dx - \int_2^6 f(x)\,dx$$

The latter integral expression is the area of the region bounded by $y = f(x)$, $x = 0$, and $x = 6$.

348. False. First, apply interval additivity to break the integral into precisely those pieces for which the intervals of integration correspond to the disjoint regions $R_1, \ldots, R_5$, as follows:

$$\int_{-2}^8 f(x)\,dx = \int_{-2}^0 f(x)\,dx + \int_0^2 f(x)\,dx + \int_2^4 f(x)\,dx + \int_4^6 f(x)\,dx + \int_6^8 f(x)\,dx$$

Now, each of the integrals on the right side where the graph of f is above the x-axis is equal to the area of that region; this applies to regions R_1 and R_2. However, each of the integrals where the graph of f is *below* the x-axis is equal to -1 times the area of that region. Therefore, the correct statement would be:

$$\int_{-2}^8 f(x)\,dx = \text{Area}\left(R_1\right) + \text{Area}\left(R_2\right) - \text{Area}\left(R_3\right) - \text{Area}\left(R_4\right) - \text{Area}\left(R_5\right)$$

349. False. Note that $\int_4^6 f(x)\,dx < 0$ because the graph of f is strictly below the x-axis on this interval. But Area (R_2) is necessarily positive, being an area. Thus, the two quantities cannot be equal. However, a true statement relating them would be $\int_4^6 f(x)\,dx = -\text{Area}\left(R_2\right)$.

350. True. Note that the region formed by pasting together R_1 and R_2 is congruent to the region formed by pasting together R_4 and R_5. As such, they have the same area. The area of the former region is given by $\int_{-2}^2 f(x)\,dx$ and the area of the latter region is given by $-\int_4^8 f(x)\,dx$ because the graph of f is below the x-axis. Thus, the given equality is true.

351. First, sketch the region bounded by $y = \cos x$, $x = \frac{\pi}{2}$, $x = \pi$, and the x-axis, as follows:

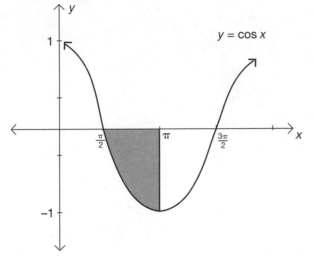

Looking back at the region, we see that if we use vertical rectangles to generate the integral, the interval of integration should be $\left[\frac{\pi}{2}, \pi \right]$ and the integrand is obtained by subtracting the top curve minus the bottom curve, which yields $0 - \cos x$. Hence, the area of the region is given by $\int_{\frac{\pi}{2}}^{\pi} [0 - \cos x] dx$. This integral is easily computed, as follows:

$$\int_{\frac{\pi}{2}}^{\pi} [0 - \cos x] dx = -\sin x \Big|_{\frac{\pi}{2}}^{\pi} = -\left[\sin \pi - \sin \frac{\pi}{2} \right] = -[0 - 1] = 1$$

So, we conclude that the area of the region is 1 unit^2.

352. First, sketch the region bounded by $y = e^{2x}$, $x = -\ln 3$, $x = \ln 2$, and the x-axis, as follows:

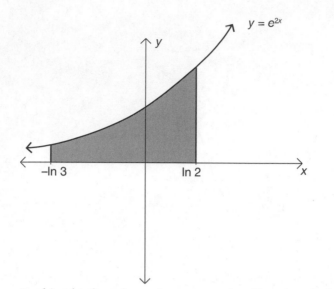

Looking back at the region, we see that if we use vertical rectangles to generate the integral, the interval of integration should be $[-\ln 3, \ln 2]$ and the integrand is obtained by subtracting the top curve minus the bottom curve, which yields $e^{2x} - 0$. Hence, the area of the region is given by $\int_{-\ln 3}^{\ln 2} \left[e^{2x} - 0 \right] dx$. In order to compute this integral, make the following substitution:

$u = 2x$

$du = 2dx \implies \frac{1}{2} du = dx$

Applying this substitution in the integrand and computing the resulting indefinite integral yields

$\int e^{2x} \, dx = \frac{1}{2} \int e^u \, du = \frac{1}{2} e^u$

Finally, rewrite the final expression in terms of the original variable x by resubstituting $u = 2x$ to obtain $\int e^{2x} \, dx = \frac{1}{2} e^{2x}$. Finally, evaluate this indefinite integral at the limits of integration and use the log rules to simplify the result, as follows:

$\int_{-\ln 3}^{\ln 2} e^{2x} \, dx = \frac{1}{2} e^{2x} \Big|_{-\ln 3}^{\ln 2} = \frac{1}{2} \left[e^{2\ln 2} - e^{-2\ln 3} \right] = \frac{1}{2} \left[e^{\ln 2^2} - e^{\ln 3^{-2}} \right] = \frac{1}{2} \left[2^2 - 3^{-2} \right]$

$= \frac{35}{18}$

So, we conclude that the area of the region is $\frac{35}{18}$ units^2.

353. First, sketch the region between $y = -x^3$ and $y = x^2$ in quadrant II, as follows:

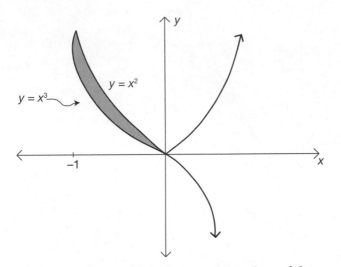

The x-coordinates of the intersection points of these two curves are obtained by equating the expressions for the two curves and solving for x, as follows:

$-x^3 = x^2$

$x^3 + x^2 = 0$

$x^2(x + 1) = 0$

$x = 0, -1$

Looking back at the region, we see that if we use vertical rectangles to generate the integral, the interval of integration should be $[-1, 0]$ and the integrand is obtained by subtracting the top curve minus the bottom curve, which yields $x^2 - (-x^3)$. Hence, the area of the region is given by $\int_{-1}^{0} \left[x^2 - \left(-x^3 \right) \right] dx$. This integral is easily computed, as follows:

$$\int_{-1}^{0} \left[x^2 - \left(-x^3 \right) \right] dx = \int_{-1}^{0} \left[x^2 + x^3 \right] dx = \frac{1}{3}x^3 + \frac{1}{4}x^4 \Big|_{-1}^{0}$$

$$= 0 - \left[\frac{1}{3}(-1)^3 + \frac{1}{4}(-1)^4 \right] = -\left[-\frac{1}{3} + \frac{1}{4} \right] = \frac{1}{12}$$

So, we conclude that the area of the region is $\frac{1}{12}$ unit^2.

354. First, sketch the region bounded by $y = \ln(ex)$, $y = \ln x$, $x = 1$, and $x = e$, as follows:

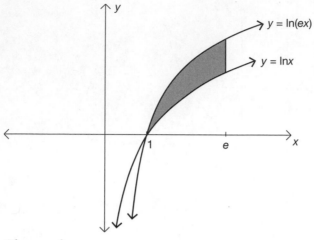

Observe that

$$\int_1^e [\ln(ex) - \ln x]\,dx = \int_1^e \left[\underbrace{\ln(e)}_{=1} + \cancel{\ln(x)} - \cancel{\ln(x)}\right]dx = \int_1^e 1\,dx = x\Big|_1^e = e - 1$$

so the area of the region is $(e - 1)$ units^2.

355. First, sketch the region bounded by $y = 3\sin x$, $y = \sin x$, $x = 0$, and $x = 2\pi$, as follows:

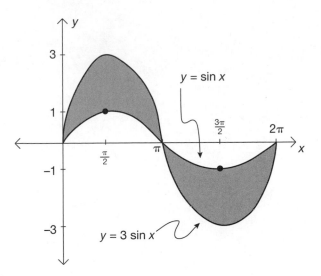

Looking back at the region, we see that if we use vertical rectangles to generate the integral, the interval of integration should be $[0, 2\pi]$ and the integrand is obtained by subtracting the top curve minus the bottom curve. However, these curves change at $x = \pi$, so we apply interval additivity and split the region into two disjoint pieces. Doing so yields the following integral expression for the area of the region:

$$\int_0^\pi [3\sin x - \sin x]\,dx + \int_\pi^{2\pi} [\sin x - 3\sin x]\,dx$$

This integral is computed as follows:

$$\int_0^\pi [3\sin x - \sin x]\,dx + \int_\pi^{2\pi} [\sin x - 3\sin x]\,dx = 2\int_0^\pi \sin x\,dx - 2\int_\pi^{2\pi} \sin x\,dx$$

$$= -2\cos x\big|_0^\pi + 2\cos x\big|_\pi^{2\pi}$$

$$= -2(\cos\pi - \cos 0)$$

$$\quad + 2(\cos 2\pi - \cos\pi)$$

$$= -2(-1-1) + 2(1-(-1))$$

$$= 8$$

So, the area of the region is 8 units^2.

Other Applications of Integration Problems

The integral arises in many applications that span different areas, including engineering, biology, chemistry, physics, and economics. We focus in this section on the following applications: volumes of solids of revolution, length of planar curves, average value, and centers of mass.

Volumes of Solids of Revolution

Consider a region in the Cartesian plane of the type displayed in the following figure:

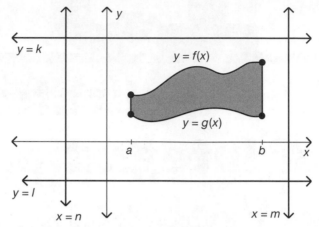

Imagine that the region is revolved 360 degrees around any one of the four horizontal or vertical lines parallel to the *x*- or *y*-axes, as displayed in the diagram; the line is treated as a central pivot around which the region is revolved. Doing so generates a solid of revolution whose volume we would like to compute. There are two main approaches to computing such volumes, namely the methods of washers and of cylindrical shells. (If you are familiar with using disks, note that this is a special case of the method of washers in which there is no hole bored through the center of the solid.) The following formulas are used to find the volumes indicated:

Description of Solid	Integral Used to Compute Volume
1. Revolve region around $y = k$	(Washers) $\pi \int_a^b \left[\underbrace{(k - g(x))^2}_{\text{big radius}} - \underbrace{(k - f(x))^2}_{\text{small radius}} \right] dx$
2. Revolve region around $y = l$	(Washers) $\pi \int_a^b \left[\underbrace{(f(x) - l)^2}_{\text{big radius}} - \underbrace{(g(x) - l)^2}_{\text{small radius}} \right] dx$
3. Revolve region around $x = n$	(Cylindrical shells) $2\pi \int_a^b \left[\underbrace{(x - n)}_{\text{radius}} \underbrace{(f(x) - g(x))}_{\text{height}} \right] dx$
4. Revolve region around $x = m$	(Cylindrical shells) $2\pi \int_a^b \left[\underbrace{(m - x)}_{\text{radius}} \underbrace{(f(x) - g(x))}_{\text{height}} \right] dx$

As with area problems, the region might be defined by boundaries that are more conveniently viewed as functions of y (instead of x). In such case, the interval of integration is taken along the y-axis, and the radii used in the washer technique and the heights used in the cylindrical shell technique are formed horizontally (rather than vertically as in the preceding scenario). Similar formulas can be developed for such a region in a completely analogous manner.

Length of a Planar Curve

The length of a smooth curve described by a differentiable function $y = f(x)$, $a \le x \le b$, is given by the integral $\int_a^b \sqrt{1 + \left(f'(x)\right)^2}\,dx$. Sometimes, a curve must be broken into disjoint pieces and this formula applied to each piece individually in order to compute its length; this situation arises, in particular, when the curve is not differentiable at a finite number of points.

Average Value of a Function

The average value of the function $y = f(x)$ on the interval $[a,b]$ is given by the integral $\frac{1}{b-a}\int_a^b f(x)\,dx$.

Centers of Mass

The center of mass $\left(\overline{x}, \overline{y}\right)$ of the planar region illustrated in the diagram provided in the earlier brief section on volumes is given by:

$$\overline{x} = \frac{\int_a^b x\,f(x)\,dx}{\text{area of region}}, \quad \overline{y} = \frac{\frac{1}{2}\int_a^b [f(x)]^2\,dx}{\text{area of region}}$$

Questions

For Questions 356 through 359, use the following region:

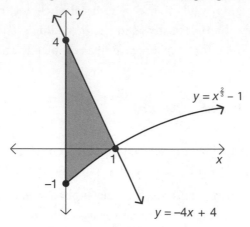

356. Which of the following can be used to compute the volume of the solid obtained by revolving the shaded region around the y-axis using the *method of washers?*

a. $\pi \int_0^1 x\left[(-4x+4)-\left(x^{\frac{2}{3}}-1\right)\right] dx$

b. $2\pi \int_0^1 x\left[(-4x+4)-\left(x^{\frac{2}{3}}-1\right)\right] dx$

c. $\pi \int_{-1}^0 \left(x^{\frac{2}{3}}-1\right)^2 dx + \pi \int_0^4 (-4x+4)^2 dx$

d. $\pi \int_0^4 \left(1-\frac{1}{4}y\right)^2 dy + \pi \int_{-1}^0 (y+1)^3 dy$

357. Which of the following can be used to compute the volume of the solid obtained by revolving the shaded region around the y-axis using the *method of cylindrical shells?*

a. $\pi \int_0^1 x\left[(-4x+4)-\left(x^{\frac{2}{3}}-1\right)\right] dx$

b. $2\pi \int_0^1 x\left[(-4x+4)-\left(x^{\frac{2}{3}}-1\right)\right] dx$

c. $\pi \int_{-1}^0 \left(x^{\frac{2}{3}}-1\right)^2 dx + \pi \int_0^4 (-4x+4)^2 dx$

d. $\pi \int_0^4 \left(1-\frac{1}{4}y\right)^2 dy + \pi \int_{-1}^0 (y+1)^3 dy$

358. Which of the following can be used to compute the volume of the solid obtained by revolving the shaded region around the line $y = -2$ using the *method of washers?*

a. $\pi\int_0^1\left[\left((-4x+4)-(-2)\right)^2-\left(\left(x^{\frac{2}{3}}-1\right)-(-2)\right)^2\right]dx$

b. $\pi\int_{-2}^4\left[(-4x+4)^2-\left(x^{\frac{2}{3}}-1\right)^2\right]dx$

c. $\pi\int_{-2}^4\left[(-4x+4)-\left(x^{\frac{2}{3}}-1\right)\right]^2 dx$

d. none of the above

359. Which of the following can be used to compute the volume of the solid obtained by revolving the shaded region around the line $x = 3$ using the *method of cylindrical shells?*

a. $2\pi\int_0^3 x\left[(-4x+4)-\left(x^{\frac{2}{3}}-1\right)\right]dx$

b. $2\pi\int_0^3 (x-3)\left[(-4x+4)-\left(x^{\frac{2}{3}}-1\right)\right]dx$

c. $2\pi\int_0^1 (3-x)\left[(-4x+4)-\left(x^{\frac{2}{3}}-1\right)\right]dx$

d. $2\pi\int_0^3 (3-x)\left[(-4x+4)-\left(x^{\frac{2}{3}}-1\right)\right]dx$

For Questions 360 and 361, use the following region:

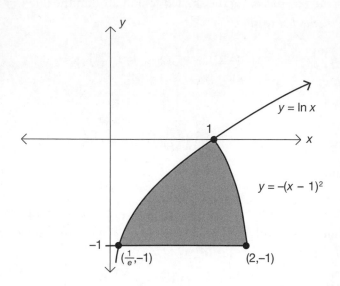

360. Which of the following can be used to compute the volume of the solid obtained by revolving the shaded region around the *x*-axis?

a. $\pi \int_{\frac{1}{e}}^{1} \left[(0-(-1))^2 - (0-\ln x)^2 \right] dx$

$+ \pi \int_{1}^{2} \left[(0-(-1))^2 - \left(0-(-(x-1)^2)\right)^2 \right] dx$

b. $2\pi \int_{-1}^{0} -y\left[\left(1+\sqrt{-y}\right) - \left(e^y\right) \right] dy$

c. both **a** and **b**

d. neither **a** nor **b**

361. Which of the following can be used to compute the volume of the solid obtained by revolving the shaded region around the *y*-axis?

a. $2\pi \int_{\frac{1}{e}}^{1} x(\ln x + 1)dx + 2\pi \int_{1}^{2} \left(1-(x-1)^2\right)dx$

b. $\pi \int_{-1}^{0} \left[\left(-(x-1)^2\right)^2 - (\ln x)^2 \right] dx$

c. both **a** and **b**

d. neither **a** nor **b**

362. Use the *method of cylindrical shells* to determine the volume of the solid obtained by revolving the region bounded by $y = x^2$ and $y = -x$ in quadrant II around the y-axis.

363. Use the *method of washers* to determine the volume of the solid obtained by revolving the region bounded by $y = x^2$ and $y = -x$ in quadrant II around the y-axis.

364. Compute the length of the portion of the curve $y = x^{\frac{3}{2}}$ starting at $x = 1$ and ending at $x = 6$.

365. The length of the portion of the curve $y = \ln x$ starting at $x = 1$ and ending at $x = e$ can be computed using which of the following?

a. $\int_1^e \sqrt{1 + (\ln x)^2}\, dx$

b. $\int_1^e \sqrt{1 + \ln x}\, dx$

c. $\int_1^e \sqrt{1 + \frac{1}{x^2}}\, dx$

d. $\int_1^e \sqrt{1 + \frac{1}{x}}\, dx$

366. The length of the portion of the curve $y = \ln(\cos x)$ starting at $x = \frac{\pi}{4}$ and ending at $x = \frac{\pi}{3}$ is equal to which of the following?

a. $\ln\left[\left(\frac{3}{\sqrt{3}}\right)\left(\frac{2}{\sqrt{2}} + 1\right)\right]$ units

b. $\ln\left(\frac{\sqrt{2} + 6}{2\sqrt{2} + \sqrt{6}}\right)$ units

c. $\ln\left[\left(2 + \sqrt{3}\right)\left(\frac{2}{\sqrt{2}} + 1\right)\right]$ units

d. $\ln\left(\frac{2\sqrt{2} + \sqrt{6}}{\sqrt{2} + 2}\right)$ units

367. The perimeter of the region bounded by $y = x^2$ and $y = -x$ in quadrant II can be computed using which of the following?

a. $\int_{-1}^{0} \sqrt{1-x}\, dx + \int_{-1}^{0} \sqrt{1+x^2}\, dx$

b. $\sqrt{2} + \int_{-1}^{0} \sqrt{1+(2x)^2}\, dx$

c. $\int_{-1}^{0} \sqrt{(-1)^2 + (2x)^2}\, dx$

d. $\int_{-1}^{0} \sqrt{-1+2x}\, dx$

368. The length of the portion of the curve $y = \int_{2}^{x} \sqrt{9t^2 - 1}\, dt$ starting at $x = 2$ and ending at $x = 3$ is equal to which of the following?

a. $\sqrt{57}$ units

b. 3 units

c. 7.5 units

d. none of the above

369. Set up, but do not evaluate, an integral expression that can be used to compute the volume of the solid obtained by revolving the region bounded by the piecewise-defined function $f(x) = \begin{cases} x^2 + 1, & 0 \le x \le 3, \\ -4x + 24, & 3 < x \le 6 \end{cases}$ and the x-axis around the y-axis using the *method of cylindrical shells.*

370. Set up, but do not evaluate, an integral expression that can be used to compute the volume of the solid obtained by revolving the region bounded by the piecewise-defined function $f(x) = \begin{cases} x^2 + 1, & 0 \le x \le 3, \\ -4x + 24, & 3 < x \le 6 \end{cases}$ and the x-axis around the x-axis using the *method of washers.*

371. The average value of the function $f(x) = x^3 \ln x$ on the interval $(1,3)$ is equal to which of the following?

    a. $\frac{81}{16}\ln 81 - 5$

    b. $\frac{81\ \ln 81 - 80}{32}$

    c. $\frac{13}{2}$

    d. $\frac{13}{4}$

372. Determine the average value of $f(x) = \sin 3x$ on the interval $\left(\frac{\pi}{12}, \frac{\pi}{9}\right)$.

373. Determine the average value of $f(x) = e^{-x}$ on the interval $(\ln 3, \ln 9)$.

374. Determine the average value of $f(x) = x^2 \sqrt[3]{x}$ on the interval $(1,8)$.

375. Determine the average value of $f(x) = x^3$ on the interval $(-2,3)$.

376. Determine the center of mass of the following region:

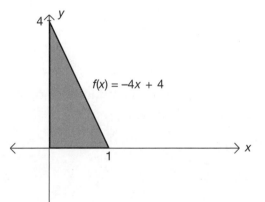

$f(x) = -4x + 4$

377. Determine the center of mass of the following region:

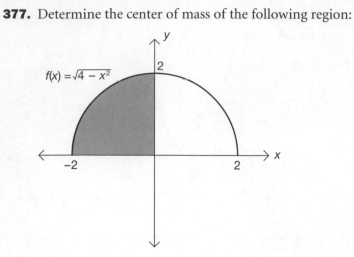

378. Which of the following integrals represents the volume of a right circular cylinder with base radius R and height $2H$?

a. $\pi \displaystyle\int_{-H}^{H} R^2 \, dx$

b. $2\pi \displaystyle\int_{-H}^{H} Rx \, dx$

c. $\pi \displaystyle\int_{0}^{R} (2H)^2 \, dx$

d. $2\pi \displaystyle\int_{0}^{R} 2Hx \, dx$

379. Which of the following integrals represents the volume of a right circular cone with base radius R and height $2H$?

a. $\pi \displaystyle\int_{0}^{R} \left[R^2 - \left(\frac{2H}{R} x\right)^2 \right] dx$

b. $\pi \displaystyle\int_{0}^{R} x \left[2H - \left(\frac{2H}{R}\right) x \right] dx$

c. $\pi \displaystyle\int_{0}^{2H} \left(\frac{R^2}{4H^2} \right) y^2 \, dy$

d. $\pi \displaystyle\int_{0}^{R} \left[2H - \left(\frac{2H}{R} x\right) \right]^2 dx$

380. Which of the following integrals represents the volume of a sphere of radius R?

a. $\pi \int_{-R}^{R} \left(R^2 - x^2 \right) dx$

b. $\pi \int_{0}^{R} \left(R^2 - x^2 \right) dx$

c. $2\pi \int_{0}^{R} x\sqrt{R^2 - x^2}\, dx$

d. $4\pi \int_{0}^{R} \sqrt{R^2 - y^2}\, dy$

Answers

356. d. Since it is specified that the method of washers is to be used to compute the volume of the solid obtained by rotating the region around the y-axis, we know that the rectangles used to form the radii must be horizontal (since they are taken to be perpendicular to the axis of rotation when using washers), so the interval of integration is taken along the y-axis and the functions describing the boundaries of the region must be solved for x in terms of y. Hence, the interval of integration is $[-1,4]$, and the boundary curves (solved for x) are given by $x = (y+1)^{\frac{3}{2}}$, $x = 1 - \frac{1}{4}y$, and the y-axis. Note that the curves used to form the radii change at $y = 0$; therefore, we split the region into two disjoint pieces, revolve each around the y-axis separately to form two individual solids that when pasted together produce the entire solid, and compute the volume of these two solids using individual integrals. Doing so and applying the formula for using washers yields

$$\pi\int_0^4 \left(1 - \tfrac{1}{4}y\right)^2 dy + \pi\int_{-1}^0 \left(y+1\right)^3 dy \,.$$

357. b. Since it is specified that the method of cylindrical shells is to be used to compute the volume of the solid obtained by rotating the region around the y-axis, we know that the rectangles used to form the height must be vertical (since they are taken to be parallel to the axis of rotation when using cylindrical shells), so the interval of integration is taken along the x-axis, and the functions describing the boundaries of the region must be solved for y in terms of x. Hence, the interval of integration is $[0,1]$, and the boundary curves (solved for y) are given by $y = x^{\frac{2}{3}} - 1$, $y = -4x + 4$, and the x-axis. Note that the curves used to form the heights do not change in the interval. Therefore, a single integral can be used to compute the volume in this case. Note that the radius is $(x - 0)$ and the height is the top curve minus the bottom curve. Thus, applying the formula for using cylindrical shells yields

$$2\pi\int_0^1 x\left[(-4x+4) - \left(x^{\frac{2}{3}} - 1\right)\right]dx \,.$$

358. a. Since it is specified that the method of washers is to be used to compute the volume of the solid obtained by rotating the region around the line $y = -2$, which is parallel to the x-axis, we know that the rectangles used to form the radii must be vertical (since they are taken to be perpendicular to the axis of rotation when using washers), so the interval of integration is taken along the x-axis, and the functions describing the boundaries of the region must be solved for y in terms of x. As such, the interval of integration is $[0,1]$, and the boundary curves (solved for x) are given by $y = x^{\frac{2}{3}} - 1, y = -4x + 4$, and the x-axis. Note that the curves used to form the heights do not change in the interval. Therefore, a single integral can be used to compute the volume in this case. Applying the formula for using washers then yields $\pi \int_0^1 \left[\left((-4x + 4) - (-2) \right)^2 - \left(\left(x^{\frac{2}{3}} - 1 \right) - (-2) \right)^2 \right] dx$.

359. c. Since it is specified that the method of cylindrical shells is to be used to compute the volume of the solid obtained by rotating the region around the line $x = 3$, which is parallel to the y-axis, we know that the rectangles used to form the height must be vertical (since they are taken to be parallel to the axis of rotation when using cylindrical shells), so the interval of integration is taken along the x-axis, and the functions describing the boundaries of the region must be solved for y in terms of x. Hence, the interval of integration is $[0,1]$, and the boundary curves (solved for y) are given by $y = x^{\frac{2}{3}} - 1, y = -4x + 4$, and the x-axis. Note that the curves used to form the heights do not change in the interval. Therefore, a single integral can be used to compute the volume in this case. Note that the radius is $(3 - x)$ and the height is the top curve minus the bottom curve. Thus, applying the formula for using cylindrical shells yields $2\pi \int_0^1 (3 - x) \left[(-4x + 4) - \left(x^{\frac{2}{3}} - 1 \right) \right] dx$.

360. **c.** We can compute the volume of the solid using both the method of washers and the method of cylindrical shells.

First, using washers, since the region is being rotated around the *x*-axis, we know that the rectangles used to form the radii must be vertical (since they are taken to be perpendicular to the axis of rotation when using washers), so the interval of integration is taken along the *x*-axis and the functions describing the boundaries of the region must be solved for *y* in terms of *x*. Thus, the interval of integration is $[\frac{1}{e},2]$. Note that the curves used to form the radii change at *x* = 1; therefore, we split the region into two disjoint pieces, revolve each around the *y*-axis separately to form two individual solids that when pasted together produce the entire solid, and compute the volume of these two solids using individual integrals. Doing so and applying the formula for using washers yields

$$\pi\int_{\frac{1}{e}}^{1}\left[(0-(-1))^2-(0-\ln x)^2\right]dx+\pi\int_{1}^{2}\left[(0-(-1))^2-\left(0-\left(-(x-1)^2\right)\right)^2\right]dx.$$

Next, using cylindrical shells, since the region is being rotated around the *x*-axis, we know that the rectangles used to form the height must be horizontal (since they are taken to be parallel to the axis of rotation when using cylindrical shells), so the interval of integration is taken along the *y*-axis, and the functions describing the boundaries of the region must be solved for *x* in terms of *y*. Thus, the interval of integration is $[-1,0]$, and the boundary curves (solved for *x*) are given by $x = e^y$ and $x = 1+\sqrt{-y}$. Note that the curves used to form the heights do not change in the interval, so a single integral can be used to compute the volume in this case. Note that the radius is $(0 - y)$, and the height is the right curve minus the left curve. Doing the computation and applying the formula for using cylindrical shells yields

$$2\pi\int_{-1}^{0}-y\left[\left(1+\sqrt{-y}\right)-\left(e^y\right)\right]dy.$$

361. **a.** Note that since the region is being revolved around the y-axis, the integral expression obtained using washers would necessarily be obtained using horizontal rectangles, and the interval of integration would need to be $[-1,0]$. The integral in choice **b** nearly satisfies these conditions, with the exception that the expressions used for the radii should have been solved for x in terms of y. So, the integral in choice **b** does *not* yield the volume of the solid. If the method of cylindrical shells is used instead, the integral expression should be obtained using vertical rectangles, and the interval of integration should be $\left[\frac{1}{e},2\right]$. Since the radius is $(x-0)$ and the height is the top curve minus the bottom curve (which is given by a single expression throughout the entire interval), using the formula for cylindrical shells yields the following integral for the volume of the solid:

$$2\pi\int_{\frac{1}{e}}^{1}x(\ln x+1)\,dx + 2\pi\int_{1}^{2}\left(1-(x-1)^2\right)dx$$

362. First, sketch the region bounded by $y = x^2$ and $y = -x$ in quadrant II, as follows:

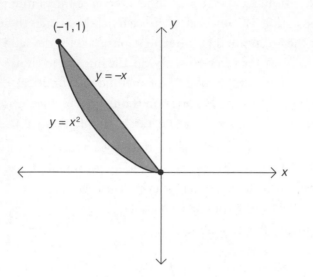

The x-coordinates of the points of intersection of these two curves are obtained by equating the expressions for the two curves and solving for x, as follows:

$x^2 = -x$

$x^2 + x = 0$

$x(x + 1) = 0$

$x = 0, -1$

Since the region is being revolved around the y-axis and the method of cylindrical shells is specified to be used, the integral expression must be formed using vertical rectangles. Hence, the interval of integration is $[-1, 0]$, the radius is $(0 - x)$, and the height is the top curve minus the bottom curve (which is given by a single expression for the entire interval). Thus, the volume of the solid is computed as follows:

$$2\pi \int_{-1}^{0} (0 - x)(-x - x^2) \, dx = 2\pi \int_{-1}^{0} (x^2 + x^3) \, dx$$

$$= 2\pi \left[\tfrac{1}{3}x^3 + \tfrac{1}{4}x^4 \right]\Big|_{-1}^{0}$$

$$= 2\pi \left[0 - \left(\tfrac{1}{3}(-1)^3 + \tfrac{1}{4}(-1)^4 \right) \right]$$

$$= 2\pi \left(\tfrac{1}{12} \right) = \tfrac{\pi}{6} \text{ units}^3$$

363. First, sketch the region bounded by $y = x^2$ and $y = -x$ in quadrant II, as follows:

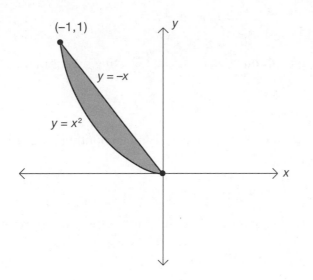

The x-coordinates of the points of intersection of these two curves were found in Question 362 to be $x = 0, -1$. The corresponding y-values of these points are 0 and 1, respectively. Since the region is being revolved around the y-axis and the method of washers is specified to be used, the integral expression must be formed using horizontal rectangles. Hence, the interval of integration is $[0,1]$, and the radii are formed by subtracting each curve minus the y-axis (i.e., $x = 0$); the expressions used for the radii remain the same throughout the entire interval.

Thus, the volume of the solid is computed, as follows:

$$\pi \int_0^1 \left[\left(0 - \sqrt{y}\right)^2 - \left(0 - (-y)\right)^2 \right] dy = \pi \int_0^1 \left[y - y^2 \right] dy$$

$$= \pi \left[\tfrac{1}{2} y^2 - \tfrac{1}{3} y^3 \right] \Big|_0^1$$

$$= \pi \left[\left(\tfrac{1}{2}(1) - \tfrac{1}{3}(1) \right) - 0 \right]$$

$$= \tfrac{\pi}{6} \text{ units}^3$$

364. Since $f'(x) = \frac{3}{2}x^{\frac{1}{2}}$, applying the length formula $\int_a^b \sqrt{1 + \left(f'(x)\right)^2}\, dx$ yields the following:

Length $= \int_1^6 \sqrt{1 + \left(\frac{3}{2}x^{\frac{1}{2}}\right)^2}\, dx = \int_1^6 \sqrt{1 + \frac{9}{4}x}\, dx$

In order to compute this integral, make the following substitution:

$u = 1 + \frac{9}{4}x$

$du = \frac{9}{4}dx \implies \frac{4}{9}du = dx$

Applying this substitution in the integrand and computing the resulting indefinite integral yields

$\int \sqrt{1 + \frac{9}{4}x}\, dx = \frac{4}{9}\int \sqrt{u}\, du = \frac{4}{9}\int u^{\frac{1}{2}}\, du = \frac{8}{27}u^{\frac{3}{2}}$

Now, rewrite the final expression above in terms of the original variable x by resubstituting $u = 1 + \frac{9}{4}x$ to obtain $\int \sqrt{1 + \left(\frac{3}{2}x^{\frac{1}{2}}\right)^2}\, dx = \frac{8}{27}\left(1 + \frac{9}{4}x\right)^{\frac{3}{2}}$.

Finally, evaluate this indefinite integral at the limits of integration to simplify the result, as follows:

$\int_1^6 \sqrt{1 + \left(\frac{3}{2}x^{\frac{1}{2}}\right)^2}\, dx = \frac{8}{27}\left(1 + \frac{9}{4}x\right)^{\frac{3}{2}}\Bigg|_1^6 = \frac{8}{27}\left[\left(1 + \frac{9}{4}(6)\right)^{\frac{3}{2}} - \left(1 + \frac{9}{4}(1)\right)^{\frac{3}{2}}\right]$

$= \frac{8}{27}\left[\left(1 + \frac{27}{2}\right)^{\frac{3}{2}} - \left(1 + \frac{9}{4}\right)^{\frac{3}{2}}\right] = \frac{8}{27}\left[\left(\frac{29}{2}\right)^{\frac{3}{2}} - \left(\frac{13}{4}\right)^{\frac{3}{2}}\right]$ units

So, we conclude that the length of the curve is $\frac{8}{27}\left[\left(\frac{29}{2}\right)^{\frac{3}{2}} - \left(\frac{13}{4}\right)^{\frac{3}{2}}\right]$ units.

365. c. Since $\frac{dy}{dx} = \frac{1}{x}$, we apply the length formula $\int_a^b \sqrt{1 + \left(f'(x)\right)^2}\, dx$ to conclude that the length of the portion of the curve $y = \ln x$ starting at $x = 1$ and ending at $x = e$ can be computed using the integral $\int_1^e \sqrt{1 + \frac{1}{x^2}}\, dx$.

366. d. Since $\frac{dy}{dx} = \frac{\sin x}{\cos x} = \tan x$, applying the length formula

$\int_a^b \sqrt{1 + \left(f'(x) \right)^2} \, dx$, together with the trigonometric identity $1 + \tan^2$

$x = \sec^2 x$, yields the following:

Length $= \int_{\frac{\pi}{4}}^{\frac{\pi}{3}} \sqrt{1 + \left(\tan x \right)^2} \, dx = \int_{\frac{\pi}{4}}^{\frac{\pi}{3}} \sqrt{\sec^2 x} \, dx = \int_{\frac{\pi}{4}}^{\frac{\pi}{3}} \sec x \, dx$

In order to compute this integral, we multiply the integrand by 1 in a very special form, as follows:

$\int_{\frac{\pi}{4}}^{\frac{\pi}{3}} \sec x \, dx = \int_{\frac{\pi}{4}}^{\frac{\pi}{3}} \sec x \cdot \frac{\sec x \, + \, \tan x}{\sec x \, + \, \tan x} \, dx = \int_{\frac{\pi}{4}}^{\frac{\pi}{3}} \frac{\sec^2 x \, + \, \sec x \, \tan x}{\sec x \, + \, \tan x} \, dx$

Now, make the following substitution:

$u = \sec x + \tan x$

$du = (\sec^2 x + \sec x \tan x) dx$

Applying this substitution in the integrand and computing the resulting indefinite integral yields $\int \sec x \, dx = \int \frac{1}{u} du = \ln|u|$. Now, rewrite the final expression of the preceding equation in terms of the original variable x by resubstituting $u = \sec x + \tan x$ to obtain $\int \sec x \, dx$

$= \ln|\sec x + \tan x|$. Finally, evaluate this indefinite integral at the limits of integration and use the log rules to simplify the result, as follows:

$\int_{\frac{\pi}{4}}^{\frac{\pi}{3}} \sec x \, dx = \ln|\sec x + \tan x| \Big|_{\frac{\pi}{4}}^{\frac{\pi}{3}}$

$= \ln\left|\sec\left(\frac{\pi}{3}\right) + \tan\left(\frac{\pi}{3}\right)\right| - \ln\left|\sec\left(\frac{\pi}{4}\right) + \tan\left(\frac{\pi}{4}\right)\right|$

$= \ln\left|2 + \sqrt{3}\right| - \ln\left|\frac{2}{\sqrt{2}} + 1\right| = \ln\left|2 + \sqrt{3}\right| - \ln\left|\frac{2 + \sqrt{2}}{\sqrt{2}}\right|$

$= \ln\left(\frac{2 + \sqrt{3}}{\frac{2 + \sqrt{2}}{\sqrt{2}}}\right) = \ln\left(\frac{\sqrt{2}\left(2 + \sqrt{3}\right)}{2 + \sqrt{2}}\right) = \ln\left(\frac{2\sqrt{2} + \sqrt{6}}{2 + \sqrt{2}}\right)$ units

367. b. First, sketch the region bounded by $y = x^2$ and $y = -x$ in quadrant II, as follows:

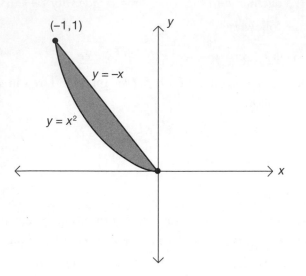

The perimeter of the region bounded by $y = x^2$ and $y = -x$ in quadrant II can be computed as the sum of two integrals: one integral for the length of the curve $y = x^2$, $-1 \le x \le 0$, and one integral for the length of the curve $y = -x$, $-1 \le x \le 0$. Applying the length formula $\int_a^b \sqrt{1 + \left(f'(x)\right)^2}\, dx$ in each of these cases yields the integral expression

$$\underbrace{\int_{-1}^0 \sqrt{1 + (-1)^2}\, dx}_{=\sqrt{2}} + \int_{-1}^0 \sqrt{1 + (2x)^2}\, dx = \sqrt{2} + \int_{-1}^0 \sqrt{1 + (2x)^2}\, dx.$$

368. c. Since the fundamental theorem of calculus implies that $f'(x) = \sqrt{9x^2 - 1}$, applying the length formula $\int_a^b \sqrt{1 + \left(f'(x)\right)^2}\, dx$ yields the following:

$$\text{Length} = \int_2^3 \sqrt{1 + \left(\sqrt{9x^2 - 1}\right)^2}\, dx = \int_2^3 \sqrt{9x^2}\, dx = \int_2^3 3x\, dx$$

$$= \tfrac{3}{2} x^2 \Big|_2^3 = \tfrac{3}{2}(9 - 4) = \tfrac{15}{2} = 7.5 \text{ units}$$

369. First, sketch the region bounded by the piecewise-defined function
$$f(x) = \begin{cases} x^2 + 1, \ 0 \le x \le 3, \\ -4x + 24, \ 3 < x \le 6 \end{cases}$$ and the x-axis, as follows:

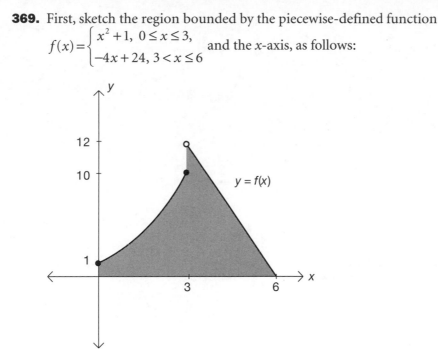

Since it is specified that the method of cylindrical shells is to be used to compute the volume of the solid obtained by revolving the region around the y-axis, the integral expression must be formed using vertical rectangles. Hence, the interval of integration is $[0,6]$, the radius is $(x - 0)$, and the height is the top curve minus the bottom curve. Given that the function f is piecewise-defined, the expression used for the top boundary changes at $x = 3$. We therefore use interval additivity to split the region into two disjoint pieces and determine the volumes of each of the two solids obtained by rotating these two pieces around the y-axis. Doing so, we see that the volume of the solid can be computed using the integral expression $2\pi \left[\int_0^3 x(x^2 + 1) \, dx + \int_3^6 x(-4x + 24) \, dx \right]$.

370. First, sketch the region bounded by the piecewise-defined function
$f(x) = \begin{cases} x^2 + 1, & 0 \le x \le 3, \\ -4x + 24, & 3 < x \le 6 \end{cases}$ and the x-axis, as follows:

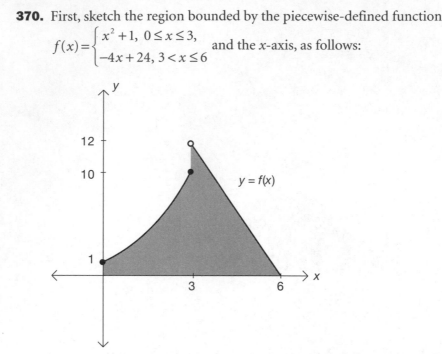

Since it is specified that the method of washers is to be used to compute the volume of the solid obtained by revolving the region around the x-axis, the integral expression must be formed using vertical rectangles. Hence, the interval of integration is $[0,6]$ and the radii are computed using the top curve minus the bottom curve. Given that the function f is piecewise-defined, the expression used for the top boundary changes at $x = 3$. We therefore use interval additivity to split the region into two disjoint pieces and determine the volumes of each of the two solids obtained by rotating these two pieces around the y-axis. Doing so, we see that the volume of the solid can be computed using the integral expression $\pi \left[\int_0^3 (x^2 + 1)^2 dx + \int_3^6 (-4x + 24)^2 dx \right]$.

371. b. The average value of the function $f(x) = x^3 \ln x$ on the interval $(1,3)$ is given by the integral expression $\frac{1}{3-1}\int_1^3 x^3 \ln x\, dx$, which is computed using the integration by parts formula $\int u\, dv = uv - \int v\, du$ with the following choices of u and v, along with their differentials:

$u = \ln x \qquad dv = x^3 dx$

$du = \frac{1}{x} dx \qquad v = \int x^3\, dx = \frac{1}{4}x^4$

Applying the integration by parts formula yields:

$$\frac{1}{3-1}\int_1^3 x^3 \ln x\, dx = \frac{1}{2}\left[\frac{1}{4}x^4 \ln x\Big|_1^3 - \int_1^3 \left(\frac{1}{4}x^4\right)\left(\frac{1}{x}\right)dx\right]$$

$$= \frac{1}{2}\left[\frac{1}{4}x^4 \ln x\Big|_1^3 - \frac{1}{4}\int_1^3 x^3\, dx\right] = \frac{1}{8}\left[x^4\left(\ln x - \frac{1}{4}\right)\right]\Big|_1^3$$

$$= \frac{1}{8}\left[3^4\left(\ln 3 - \frac{1}{4}\right) - 1^4\left(\ln 1 - \frac{1}{4}\right)\right] = \frac{1}{8}\left[81\left(\frac{4\ln 3 - 1}{4}\right) + \frac{1}{4}\right]$$

$$= \frac{1}{32}\left[81\left(\ln\left(3^4\right) - 1\right) + 1\right] = \frac{81\ \ln 81 - 80}{32}$$

372. The average value of $f(x) = \sin 3x$ on the interval $\left(\frac{\pi}{12}, \frac{\pi}{9}\right)$ is given by the

integral expression $\frac{1}{\frac{\pi}{9} - \frac{\pi}{12}} \int_{\frac{\pi}{12}}^{\frac{\pi}{9}} \sin 3x \, dx$. To compute this integral, we make

the following substitution:

$u = 3x$

$du = 3 \, dx \implies \frac{1}{3} du = dx$

Applying this substitution in the integrand and computing the resulting

indefinite integral yields $\int \sin 3x \, dx = \frac{1}{3} \int \sin u \, du = -\frac{1}{3} \cos u$. Now,

rewrite the final expression in terms of the original variable x by

resubstituting $u = 3x$ to obtain $\int \sin 3x \, dx = -\frac{1}{3} \cos 3x$. Finally, evaluate

this indefinite integral at the limits of integration and use the log rules

to simplify the result, as follows:

$$\frac{1}{\frac{\pi}{9} - \frac{\pi}{12}} \int_{\frac{\pi}{12}}^{\frac{\pi}{9}} \sin 3x \, dx = \left(\frac{36}{\pi}\right)\left(-\frac{1}{3}\right) \cos 3x \Big|_{\frac{\pi}{12}}^{\frac{\pi}{9}} = -\frac{12}{\pi}\left(\cos\left(\frac{\pi}{3}\right) - \cos\left(\frac{\pi}{4}\right)\right)$$

$$= -\frac{12}{\pi}\left(\frac{1}{2} - \frac{\sqrt{2}}{2}\right) = -\frac{6}{\pi}\left(1 - \sqrt{2}\right) = \frac{6\left(\sqrt{2} - 1\right)}{\pi}$$

373. The average value of $f(x) = e^{-x}$ on the interval $(\ln 3, \ln 9)$ is given by the

integral expression $\frac{1}{\ln 9 - \ln 3} \int_{\ln 3}^{\ln 9} e^{-x} \, dx$. Since the antiderivative of

e^{-x} is $-e^{-x}$, this integral is computed as follows:

$$\frac{1}{\ln 9 - \ln 3} \int_{\ln 3}^{\ln 9} e^{-x} \, dx = \frac{-1}{\ln 9 - \ln 3} e^{-x} \Big|_{\ln 3}^{\ln 9} = \frac{-\left(e^{-\ln 9} - e^{-\ln 3}\right)}{\ln 9 - \ln 3}$$

$$= \frac{-\left(e^{\ln 9^{-1}} - e^{\ln 3^{-1}}\right)}{\ln 9 - \ln 3} = \frac{-\left(\frac{1}{9} - \frac{1}{3}\right)}{\ln 3^2 - \ln 3} = \frac{\frac{2}{9}}{2\ln 3 - \ln 3} = \frac{2}{9\ln 3}$$

374. The average value of $f(x) = x^2 \sqrt[3]{x}$ on the interval $(1,8)$ is computed as

follows:

$$\frac{1}{8-1} \int_1^8 x^2 \sqrt[3]{x} \, dx = \frac{1}{7} \int_1^8 x^{\frac{7}{3}} \, dx = \frac{1}{7} \cdot \frac{3}{10} x^{\frac{10}{3}} \Big|_1^8 = \frac{3}{70}\left(8^{\frac{10}{3}} - 1\right) = \frac{3}{70}\left(2^{10} - 1\right) = \frac{3,069}{70}$$

375. The average value of $f(x) = x^3$ on the interval $(-2,3)$ is computed as

follows:

$$\frac{1}{3-(-2)} \int_{-2}^3 x^3 \, dx = \frac{1}{5} \cdot \frac{1}{4} x^4 \Big|_{-2}^3 = \frac{1}{20}\left(3^4 - (-2)^4\right) = \frac{65}{20} = \frac{13}{4}$$

376. The coordinates of the center of mass $(\bar{x}, \bar{y})$ are given by:

$$\bar{x} = \frac{\int_a^b x f(x)\, dx}{\text{area of region}}, \quad \bar{y} = \frac{\frac{1}{2}\int_a^b [f(x)]^2\, dx}{\text{area of region}}$$

The following computations are needed:

$$\text{Area} = \int_0^1 (-4x + 4)\, dx = -2x^2 + 4x \Big|_0^1 = 2$$

$$\int_0^1 x f(x)\, dx = \int_0^1 x(-4x + 4)\, dx = \int_0^1 (-4x^2 + 4x)\, dx = \left(-\frac{4x^3}{3} + 2x^2\right)\Big|_0^1 = \frac{2}{3}$$

$$\int_0^1 \frac{1}{2}\big(f(x)\big)^2\, dx = \int_0^1 \frac{1}{2}(-4x + 4)^2\, dx = 8\int_0^1 \left(x^2 - 2x + 1\right) dx = 8\left(\frac{x^3}{3} - x^2 + x\right)\Big|_0^1$$

$$= \frac{8}{3}$$

Thus, $\bar{x} = \frac{\frac{2}{3}}{2} = \frac{1}{3}$, $\bar{y} = \frac{\frac{8}{3}}{2} = \frac{4}{3}$, so the center of mass is $\left(\frac{1}{3}, \frac{4}{3}\right)$.

377. The coordinates of the center of $(\overline{x}, \overline{y})$ mass are given by:

$$\overline{x} = \frac{\int_a^b x f(x)\,dx}{\text{area of region}}, \quad \overline{y} = \frac{\frac{1}{2}\int_a^b [f(x)]^2\,dx}{\text{area of region}}$$

The following computations are needed:

$$\text{Area} = \tfrac{1}{4}(\text{area of circle of radius } 2) = \tfrac{1}{4}\left(\pi(2)^2\right) = \pi$$

$$\int_{-2}^0 \tfrac{1}{2}\left(f(x)\right)^2\,dx = \int_{-2}^0 \tfrac{1}{2}\left(\sqrt{4-x^2}\right)^2\,dx = \int_{-2}^0 \tfrac{1}{2}\left(4-x^2\right)dx$$

$$= \tfrac{1}{2}\left(4x - \tfrac{x^3}{3}\right)\Big|_{-2}^{0} = \tfrac{1}{2}\left[0 - \left(-8 + \tfrac{8}{3}\right)\right] = \tfrac{8}{3}$$

$$\int_{-2}^0 x f(x)\,dx = \int_{-2}^0 x\sqrt{4-x^2}\,dx$$

To compute the last integral, make the following substitution:

$$u = 4 - x^2$$

$$du = -2x\,dx \;\Rightarrow\; -\tfrac{1}{2}du = x\,dx$$

Applying this substitution in the integrand and computing the resulting indefinite integral yields $\int x\sqrt{4-x^2}\,dx = -\tfrac{1}{2}\int \sqrt{u}\,du = -\tfrac{1}{2}\cdot\tfrac{2}{3}u^{\frac{3}{2}} = -\tfrac{1}{3}u^{\frac{3}{2}}$.

Now, rewrite the final expression above in terms of the original variable x by resubstituting $u = 4 - x^2$ to obtain $\int x\sqrt{4-x^2}\,dx = -\tfrac{1}{3}\left(4-x^2\right)^{\frac{3}{2}}$.

Finally, evaluate this indefinite integral at the limits of integration and use the log rules to simplify the result, as follows:

$$\int_{-2}^0 x f(x)\,dx = \int_{-2}^0 x\sqrt{4-x^2}\,dx$$

$$= -\tfrac{1}{3}\left(4-x^2\right)^{\frac{3}{2}}\Big|_{-2}^{0} = -\tfrac{1}{3}\left[(4-0)^{\frac{3}{2}} - \left(4-(-2)^2\right)^{\frac{3}{2}}\right]$$

$$= -\tfrac{1}{3}(4-0)^{\frac{3}{2}} = -\tfrac{8}{3}$$

Thus, $\overline{x} = \dfrac{-\frac{8}{3}}{\pi} = -\tfrac{8}{3\pi}$, $\overline{y} = \dfrac{\frac{8}{3}}{\pi} = \tfrac{8}{3\pi}$, so the center of mass is $\left(-\tfrac{8}{3\pi}, \tfrac{8}{3\pi}\right)$.

378. d. Consider the region bounded by $y = H$, $y = -H$, $x = 0$, and $x = R$, shown here:

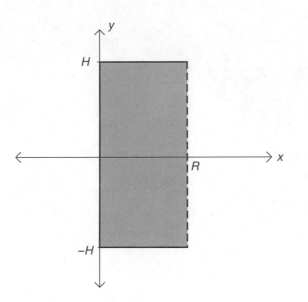

If this region is revolved around the y-axis, a right circular cylinder with base radius R and height $2H$ is formed. Using the method of cylindrical shells, the volume of this solid is given by the integral $2\pi \int_{0}^{R} 2Hx\, dx$.

379. **c.** Consider the region in quadrant I bounded by $x = 0$, $y = 2H$, and the line $y = \frac{2H}{R}x$, shown here:

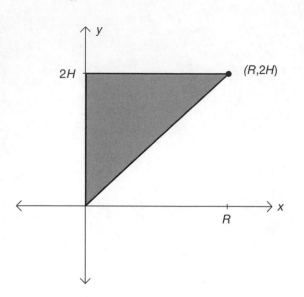

If this region is revolved around the y-axis, a right circular cone with base radius R and height $2H$ is formed. Using the method of washers, the volume of this solid is given by the integral $\pi \int_0^{2H} \left(\frac{R^2}{4H^2} \right) y^2 \, dy$.

380. a. Consider the region bounded by $y = 0$ and the semicircle whose equation is given by $y = \sqrt{R^2 - x^2}$ shown here:

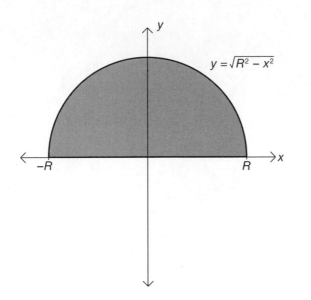

$y = \sqrt{R^2 - x^2}$

If this region is revolved around the x-axis, a sphere of radius R is formed. Using the method of washers, the volume of this solid is given by the integral $\pi \int_{-R}^{R} \left(R^2 - x^2 \right) dx$.

Differential
Equations Problems

A *differential equation,* simply put, is an equation involving one or more derivatives of some unknown function. A *solution* to a differential equation is a function that satisfies the equation in the sense that if all derivatives arising in the equation are computed and substituted back into the equation, a true statement will be the outcome. In the absence of any additional conditions being imposed along with the differential equation, a function satisfying it will contain arbitrary constants, or *parameters*; such a solution is called the *general solution* of the differential equation. Imposing additional restrictions on the solution, such as demanding it pass through a specific point or that its slope at a certain input must equal a specified value, results in a *particular solution* of the differential equation. There are numerous techniques used to solve such equations by hand, the most rudimentary of which is the method of *separating variables*.

Differential equations arise naturally in the elementary mathematical modeling of population growth and half-life of radioactive isotopes, for instance. These particular models are described by *initial-value problems*, which consist of a differential equation and initial conditions imposed on the unknown function and some of its derivatives. A *solution* to an initial-value problem is a function that satisfies both the differential equation in the sense described earlier and the initial conditions specified in the model.

Questions

381. The family of curves described by the function $y(x) = C_1 e^{2x} + C_2 e^{-3x}$, where C_1 and C_2 are arbitrary real constants, satisfies which of the following differential equations?

a. $y''(x) + y'(x) - 6y(x) = 0$

b. $y''(x) - y'(x) - 6y(x) = 0$

c. $\left(y'(x) - 2y(x)\right)\left(y'(x) + 3y(x)\right) = 0$

d. $\left(y'(x) + 2y(x)\right)\left(y'(x) - 3y(x)\right) = 0$

382. The equation of the function $y(x)$ whose graph passes through the point $(1,1)$ and whose tangent line slope at x is given by $\frac{dy}{dx} = xe^x$ is which of the following?

a. $y(x) = \frac{1}{2}x^2 e^x + 1 - \frac{e}{2}$

b. $y(x) = e^x(x-1) + 1$

c. $y(x) = e^x(x-1)$

d. $y(x) = xe^x + e^x + 1$

383. The general solution of the differential equation $\frac{dy}{dx} = \frac{x}{y}$ that is known to pass through some point above the x-axis is given by which of the following?

a. $y(x) = kx$, where k is a positive real number

b. $y(x) = \sqrt{x^2 + k}$, where k is a real number

c. $y(x) = x + k$, where k is a positive real number

d. none of the above

384. The general solution of the differential equation $\frac{dy}{dx} = \frac{x}{x+1}$ is which of the following?

a. $y(x) = x - \frac{1}{(x+1)^2} + C$, where C is a real number

b. $y(x) = x + \frac{1}{2}x^2 + C$, where C is a real number

c. $y(x) = \frac{x^2}{x^2 + x} + C$, where C is a real number

d. $y(x) = x - \ln|x+1| + C$, where C is a real number

385. The general solution of the differential equation $\frac{dx}{dy} = \frac{x^2 + 1}{\sec y}$ is which of the following?

    a. $y(x) = \arcsin(\arctan x + C)$, where C is a real number
    b. $y(x) = \arcsin(\arctan x) + C$, where C is a real number
    c. $y(x) = \arcsin(-\arctan x + C)$, where C is a real number
    d. none of the above

386. Suppose that the growth of a particular population of bacteria is described by the differential equation $\frac{dy}{dt} = 5y\left(2 - \frac{y}{100}\right)$. For what value(s) of y is the population decreasing?

387. The half-life of a certain radioactive isotope is 420 years. A sample of 40 mg is taken. Determine the mass remaining after 350 years, rounded to the nearest tenth of a milligram.

388. The half-life of a certain radioactive isotope is 420 years. A sample of 40 mg is taken. At what time, rounded to the nearest year, is the mass 15 mg?

389. A sample of a radioactive isotope has decayed to 80% of its original size, y_0, after 1.5 years. Determine its half-life to the nearest tenth of a year.

390. True or false? The general solution $y(x)$ of the differential equation $y'(x) = -\sqrt{x^2 + y^2}$ is not increasing on any subset of its domain.

391. True or false? The function $y(x) = C_1 + C_2 x + C_3 x^2 + C_4 x^3$, where C_i ($i = 1, 2, 3, 4$) is any real number, is a solution of the differential equation $y'''(x) = 0$.

392. Determine the general solution of the differential equation $\frac{x^2 y + y}{y^2 + 1} \cdot \frac{dy}{dx} = 1$.

393. Determine the general solution of the differential equation
$e^{2x}\left(\cos y+1\right)^2\frac{dx}{dy}=e^{-5x}\sin y$.

394. True or false? The differential equation $\left(\frac{dy}{dx}\right)^2+y^2=\ln\left(\frac{1}{2+y^2}\right)$ has no solution.

395. True or false? The solution of the initial-value problem $\begin{cases}\frac{dy}{dt}=y(y-2)\\y(0)=2\end{cases}$ is a constant function.

Answers

381. a. Observe that $y(x) = C_1e^{2x} - C_2e^{-3x}$ and $y''(x) = 4C_1e^{2x} + 9C_2e^{-3x}$. Substituting these functions, along with the given function $y(x) = C_1e^{2x} + C_2e^{-3x}$, into the differential equation provided in choice **a** yields

$$y''(x) + y'(x) - 6y(x) = \left(4C_1e^{2x} + 9C_2e^{-3x}\right) + \left(2C_1e^{2x} - 3C_2e^{-3x}\right)$$
$$-6\left(C_1e^{2x} + C_2e^{-3x}\right)$$
$$= \left(4C_1 + 2C_1 - 6C_1\right)e^{2x} + \left(9C_2 - 3C_2 - 6C_2\right)e^{-3x}$$
$$= 0$$

382. b. Separate variables and integrate both sides to obtain

$$dy = xe^x dx \implies \int dy = \int xe^x\, dx$$

The integral on the left side is simply y. Compute the integral on the right side using integration by parts. Let

$u = x \qquad dv = e^x dx$

$du = dx \quad v = e^x$

Hence,

$$\int xe^x\, dx = uv - \int v\, du = xe^x - \int e^x dx = xe^x - e^x + C = e^x(x-1) + C$$

So, the general solution of the differential equation is $y(x) = e^x(x-1) + C$. Now, recall that in order for the curve to pass through the point $(1,1)$, it must be the case that $y(1) = 1$. Evaluating the general solution at $x = 1$ yields $y(1) = e^1(1-1) + C = C$. Thus, we conclude that $C = 1$ by equating these two. Therefore, the particular solution we seek is $y(x) = e^x(x-1) + 1$.

383. b. Separate variables, integrate both sides, and solve for y to obtain

$y\,dy = x\,dx$

$\int y\,dy = \int x\,dx$

$\frac{1}{2}y^2 = \frac{1}{2}x^2 + C$

$y^2 = x^2 + k$

A remark is in order before proceeding. While it is technically the case that $k = 2C$, keep in mind that C is an arbitrary constant of integration. As such, multiplying C by 2 does not render it less arbitrary. So, we simply relabel the arbitrary constant as k to avoid unnecessarily cumbersome bookkeeping. Now, at this point, if we solve the last equation for y, we would obtain two possible expressions for y, namely $\pm\sqrt{x^2 + k}$. Using the fact that the graph of $y(x)$ is known to pass through some point above the x-axis prompts us to conclude that $y(x) = \sqrt{x^2 + k}$ because the expression $-\sqrt{x^2 + k}$ is always less than or equal to zero.

384. d. Separate variables and integrate both sides to obtain

$dy = \frac{x}{x+1}\,dx \;\Rightarrow\; \int dy = \int \frac{x}{x+1}\,dx$

The integral on the left-side is simply y. Compute the integral on the right side as follows:

$\int \frac{x}{x+1}\,dx = \int \frac{x+1-1}{x+1}\,dx = \int\left[\frac{x+1}{x+1} - \frac{1}{x+1}\right]dx = \int\left[1 - \frac{1}{x+1}\right]dx$

$= x - \ln|x+1| + C$

Thus, $y(x) = x - \ln|x+1| + C$.

385. a. Separate variables, integrate both sides, and solve for y to obtain

$\frac{dx}{dy} = \frac{x^2+1}{\sec y}$

$\frac{dx}{x^2+1} = \frac{dy}{\sec y} = \cos y\,dy$

$\int \frac{dx}{x^2+1} = \int \cos y\,dy$

$\arctan x + C = \sin y$

$y(x) = \arcsin(\arctan x + C)$

386. The solution of this differential equation is the function y. This function is decreasing precisely when $\frac{dy}{dt} < 0$. Thus, we must determine the nonnegative values of y for which $5y\left(2 - \frac{y}{100}\right) < 0$; we restrict our attention to nonnegative values of y because it represents a population. Note that the left side equals zero when $y = 0, 200$. The first factor is positive for all positive values of y, while the second factor is negative whenever $y > 200$. Thus, the product of the factors is negative whenever $y > 200$. We conclude that the population is decreasing for all values of $y > 200$.

387. Let $y(t)$ represent the mass of the isotope at time t, where t is measured in years. We must solve the initial-value problem

$$\begin{cases} \frac{dy}{dt} = ky \\ y(0) = 40 \end{cases}$$

First, separate variables, integrate both sides, and solve for y to obtain

$\int \frac{dy}{y} = \int k\, dt$

$\ln|y| = kt + C$

$|y(t)| = e^{kt+C} = e^C e^{kt} = \overline{C} e^{kt}$

where $\overline{C}$ is a positive real number. Since $y(t)$ represents a population in the present context, and such a quantity must be nonnegative, we can remove the absolute value and retain the same characterization of the constant $\overline{C}$. That is, the general solution is $y(t) = \overline{C} e^{kt}$.

We must determine the values of both $\overline{C}$ and k. To this end, we apply the initial condition $y(0) = 40$ in the general solution to obtain

$40 = y(0) = \overline{C} e^{k(0)} = \overline{C}$

This enables us to refine the general solution to the form $y(t) = 40e^{kt}$. Now, in order to determine k, we use the fact that the half-life is 420 years, meaning that 20 mg of the isotope would be present after 420 years. Equivalently, $y(420) = 20$. Applying this condition to the refined form of the solution yields

$40e^{k(420)} = 20$

$e^{k(420)} = \frac{1}{2}$

$k(420) = \ln\left(\frac{1}{2}\right)$

$k = \frac{1}{420}\ln\left(\frac{1}{2}\right)$

Thus, the solution of the initial-value problem is given by $y(t) = 40e^{\frac{1}{420}\ln\left(\frac{1}{2}\right)t}$.

Finally, to answer the question posed, we evaluate this function at $t = 350$ to obtain

$y(350) = 40e^{\left(\frac{1}{420}\ln\left(\frac{1}{2}\right)\right)(350)} \approx 22.5$

So, approximately 22.5 mg remain after 350 years.

388. This scenario is described by the same initial-value problem as in the previous problem. As such, the mass at time t is given by $y(t) = 40e^{\frac{1}{420}\ln\left(\frac{1}{2}\right)t}$. Answering the question posed requires that we solve the following equation for t:

$40e^{\frac{1}{420}\ln\left(\frac{1}{2}\right)t} = 15$

$e^{\frac{1}{420}\ln\left(\frac{1}{2}\right)t} = \frac{3}{8}$

$\frac{1}{420}\ln\left(\frac{1}{2}\right)t = \ln\left(\frac{3}{8}\right)$

$t = \frac{420\ln\left(\frac{3}{8}\right)}{\ln\left(\frac{1}{2}\right)} \approx 594$

So, it takes approximately 594 years for the initial mass of 40 mg to decay to a mass of size 15 mg.

389. Let $y(t)$ represent the mass of the isotope at time t, where t is measured in years. We must solve the initial-value problem

$$\begin{cases} \frac{dy}{dt} = ky \\ y(1.5) = 0.80 y_0 \end{cases}$$

First, separate variables, integrate both sides, and solve for y to obtain

$\int \frac{dy}{y} = \int k\,dt$

$\ln|y| = kt + C$

$|y(t)| = e^{kt+C} = e^C e^{kt} = \overline{C}e^{kt}$

where $\overline{C}$ is a positive real number. Since $y(t)$ represents a population in the present context and such a quantity must be nonnegative, we can remove the absolute value and retain the same characterization of the constant $\overline{C}$. That is, the general solution is $y(t) = \overline{C}e^{kt}$.

We must determine the values of both $\overline{C}$ and k. To this end, we apply the initial condition $y(0) = y_0$ in the general solution to obtain

$y_0 = y(0) = \overline{C}e^{k(0)} = \overline{C}$

This enables us to refine the general solution to the form $y(t) = y_0 e^{kt}$. Now, in order to determine k, we use the other condition provided, namely $y(1.5) = 0.80 y_0$. Applying this condition to the refined form of the solution yields

$0.80 y_0 = y_0 e^{k(1.5)}$

$0.80 = e^{k(1.5)}$

$\ln(0.80) = k(1.5)$

$k = \frac{\ln(0.80)}{1.5} = \frac{2}{3}\ln\left(\frac{4}{5}\right)$

Thus, the solution of the initial-value problem is given by $y(t) = y_0 e^{\frac{2}{3}\ln\left(\frac{4}{5}\right)t}$.

Finally, since the half-life is the time t for which $y(t) = \frac{1}{2}y_0$, we must solve the following equation for t:

$$y_0 e^{\frac{2}{3}\ln\left(\frac{4}{5}\right)t} = \frac{1}{2}y_0$$

$$e^{\frac{2}{3}\ln\left(\frac{4}{5}\right)t} = \frac{1}{2}$$

$$\frac{2}{3}\ln\left(\frac{4}{5}\right)t = \ln\left(\frac{1}{2}\right)$$

$$t = \frac{\ln\left(\frac{1}{2}\right)}{\frac{2}{3}\ln\left(\frac{4}{5}\right)} \approx 4.659$$

So, the half-life is approximately 4.7 years.

390. True. The differential equation itself implies that $\frac{dy}{dx} \leq 0$, for all values of x. Thus, the solution y is either constant or decreasing on its entire domain.

391. False. Since $y'''(x) = 6C_4$, it must be the case that $C_4 = 0$ in order for the function $y(x) = C_1 + C_2 x + C_3 x^2 + C_4 x^3$ to be a solution to the given differential equation.

392. Separate variables and then integrate both sides, as follows:

$$\frac{x^2 y + y}{y^2 + 1} \cdot \frac{dy}{dx} = 1$$

$$\frac{y\left(x^2 + 1\right)}{y^2 + 1} dy = dx$$

$$\frac{y}{y^2 + 1} dy = \frac{1}{x^2 + 1} dx$$

$$\int \frac{y}{y^2 + 1} dy = \int \frac{1}{x^2 + 1} dx$$

The integral on the right side is arctan$x + C$. Computing the integral on the left side requires a substitution. Specifically, let $u = y^2 + 1$, $du = 2y\,dy$. Observe that

$$\int \frac{y}{y^2 + 1} dy = \int \frac{\frac{1}{2} du}{u} = \frac{1}{2} \int \frac{du}{u} = \frac{1}{2}\ln|u| = \frac{1}{2}\ln\left|y^2 + 1\right| = \frac{1}{2}\ln\left(y^2 + 1\right)$$

(Note that the absolute value can be dropped on the last term in this string of equalities because $y^2 + 1 > 0$ for all values of y.) Hence, the general solution is defined implicitly as $\frac{1}{2}\ln\left(y^2 + 1\right) = \arctan x + C$. Technically, this equation can be solved explicitly for y, but it requires one to choose between a positive expression and a negative expression, which requires the use of an initial condition. Since such a condition is not specified, we solve the expression for y^2, as follows:

$$\frac{1}{2}\ln\left(y^2 + 1\right) = \arctan x + C$$

$$\ln\left(y^2 + 1\right) = 2\arctan x + 2C$$

$$y^2 + 1 = e^{2\arctan x + 2C} = \overline{C}e^{2\arctan x}$$

$$y^2 = \overline{C}e^{2\arctan x} - 1$$

where $\overline{C}$ is a positive real number.

393. Separate variables and then integrate both sides, as follows:

$$e^{2x}\left(\cos y+1\right)^2 \frac{dx}{dy}=e^{-5x}\sin y$$

$$\frac{e^{2x}}{e^{-5x}}dx=\frac{\sin y}{\left(\cos y+1\right)^2}dy$$

$$\int e^{7x}\,dx=\int \frac{\sin y}{\left(\cos y+1\right)^2}dy$$

Computing both integrals requires the use of substitutions. For the integral on the left side, let $u=7x$, $du=7dx$. Observe that

$$\int e^{7x}\,dx=\int e^u\,\tfrac{1}{7}du=\tfrac{1}{7}\int e^u\,du=\tfrac{1}{7}e^u+C=\tfrac{1}{7}e^{7x}+C$$

For the integral on the right side, let $w=\cos y+1$, $dw=-\sin y\,dy$. Observe that

$$\int \frac{\sin y}{\left(\cos y+1\right)^2}dy=\int \frac{-dw}{w^2}=-\int w^{-2}\,dw=-\left(-w^{-1}\right)=\tfrac{1}{w}=\frac{1}{\cos y+1}$$

Hence, the general solution is defined implicitly as $\frac{1}{\cos y+1}=\tfrac{1}{7}e^{7x}+C$, or equivalently as $1=\left(\tfrac{1}{7}e^{7x}+C\right)\left(\cos y+1\right)$.

394. True. Observe that $0<\frac{1}{2+y^2}\le \tfrac{1}{2}$, for all values of y. Thus, $\ln\left(\frac{1}{2+y^2}\right)<0$, for all values of y. Moreover, the given differential equation is equivalent to

$$\left(\tfrac{dy}{dx}\right)^2=-y^2+\ln\left(\tfrac{1}{2+y^2}\right)$$

But the left side is always nonnegative while the right side is negative, for all values of y. Consequently, there cannot exist a function y that satisfies this equation. Hence, the differential equation has no solution.

395. True. We begin by constructing the general solution of the differential equation. To this end, separate variables and then integrate both sides to obtain:

$$\frac{dy}{y(y-2)} = dt$$

$$\int \frac{dy}{y(y-2)} = \int dt$$

The integral on the right side is simply $t + C$. Computing the integral on the left side requires that we first rewrite the integrand as an equivalent sum of two simpler fractions using partial fraction decomposition. Precisely, we must determine constants A and B such that

$$\frac{1}{y(y-2)} = \frac{A}{y} + \frac{B}{y-2}$$

Multiplying both sides of this equation by the lowest common deonominator (LCD) $y(y-2)$, and then grouping like terms, yields

$$1 = A(y-2) + By = (A+B)y - 2A$$

Equating the coefficients of terms of the same degree on both sides of this equation yields the following system of linear equations:

$$\begin{cases} A+B=0 \\ -2A=1 \end{cases}$$

To solve this system, solve the second equation for A to obtain $A = -\frac{1}{2}$.

Substituting this value into the first equation subsequently yields $B = \frac{1}{2}$. Hence, we conclude that

$$\frac{1}{y(y-2)} = \frac{-\frac{1}{2}}{y} + \frac{\frac{1}{2}}{y-2}$$

We can now integrate, as follows:

$$\int \frac{1}{y(y-2)} dy = \int \left(\frac{-\frac{1}{2}}{y} + \frac{\frac{1}{2}}{y-2} \right) dy = -\frac{1}{2} \int \frac{1}{y} dy + \frac{1}{2} \int \frac{1}{y-2} dy$$

$$= -\frac{1}{2} \ln|y| + \frac{1}{2} \ln|y-2|$$

(Technically, a substitution was used to compute the second integral in the preceding line.) Simplifying using logarithm properties yields

$$\int \frac{1}{y(y-2)} dy = \frac{1}{2} \ln \left| \frac{y-2}{y} \right|$$

Therefore, the general solution of the given differential equation is $\frac{1}{2}\ln\left|\frac{y-2}{y}\right| = t + C$.

Next, solve this equation explicitly for y, as follows:

$$\frac{1}{2}\ln\left|\frac{y-2}{y}\right| = t + C$$

$$\ln\left|\frac{y-2}{y}\right| = 2t + 2C$$

$$\left|\frac{y-2}{y}\right| = e^{2t+2C} = e^{2t}e^{2C} = Ke^{2t}$$

At this point, K represents a positive real number. However, if we remove the absolute values, then we can expand the set of tenable values for K to all nonzero real numbers. With this in mind, we continue solving the above equation for y:

$$\frac{y-2}{y} = Ke^{2t}$$

$$y - 2 = Ke^{2t}y$$

$$y - Ke^{2t}y = 2$$

$$y\left(1 - Ke^{2t}\right) = 2$$

$$y = \frac{2}{1 - Ke^{2t}}$$

So, the general solution of the differential equation is $y(t) = \frac{2}{1 - Ke^{2t}}$.
Finally, we must determine the value of K for which this solution satisfies the initial condition $y(0) = 2$. We simply substitute this into the general form of the solution to obtain

$$2 = y(0) = \frac{2}{1 - Ke^{2(0)}} = \frac{2}{1 - K}$$

$$2(1 - K) = 2$$

$$K = 0$$

Hence, we conclude that the solution of the initial-value problem is the constant function $y(t) = 2$.

17

Sequence and Infinite Series Problems

Sequences naturally arise in any formal study of approximation or limits of any kind. An *infinite sequence of real numbers* is simply an ordered arrangement of real numbers. Sequences are usually denoted by an ordered arrangement of terms separated by commas, such as:

$$x_1, x_2, \ldots, \underbrace{\{x_n}_{n\text{th term}}, \ldots n = 1, 2, 3, \ldots$$

More succinctly, in a given discussion, it is customary to write $\{x_n\}$.

How to Correctly Express a Sequence

The two most common ways to define/express a sequence are:

1. Writing out the first few terms of an ordered list to express the pattern which characterizes the sequence.
2. Determining an explicit formula for the nth term.

Examples

1. The sequence $0, \frac{1}{2}, \frac{2}{3}, \frac{3}{4}, \ldots$ can also be expressed as

   $x_n = \frac{n}{n+1}$, $n = 0, 1, 2, \ldots$,

   or equivalently as

   $y_n = \frac{n-1}{n}$, $n = 1, 2, 3, \ldots$.

2. The sequence $\frac{3}{16}, \frac{4}{25}, \frac{5}{36}, \ldots$ can also be

   expressed as $x_n = \frac{n}{(n+1)^2}$, $n = 3, 4, 5, \ldots$

   The terms of the sequences given above all seem to be heading toward some sort of target value.

Convergence of a Sequence

A real-valued sequence $\{x_n\}$ is said to be *convergent* to l if as n gets larger and larger, $|x_n - l|$ gets arbitrarily close to 0. In such case, we write $\lim_{n \to \infty} x_n = l$ (or simply $x_n \to l$). If there is no such target l for which $x_n \to l$, we say that $\{x_n\}$ is *divergent*.

Connection to a function: Let $x_n = x(n)$, $n = 1, 2, 3, \ldots$ and define a function f such that $f(n) = x_n$, $n = 1, 2, 3, \ldots$; if there exists $\lim_{n \to \infty} f(x) = l$, then there exists $\lim_{n \to \infty} x_n = l$, but not conversely. In words, if the graph of the function whose rule defines the sequence $\{x_n\}$ has a horizontal asymptote (as $x \to \infty$), then the sequence will converge to the same limit value.

Note that connecting the study of convergence of sequences to the study of limits of *functions* makes available to us the entire arsenal built when studying limits of functions to aid in computing limits of *sequences*. These include:

- some standard rules for powers and quotients
- basic algebraic tricks (e.g., rationalizing, factoring and canceling, using known trigonometric identities)

- knowledge of behavior of common graphs, together with continuity
- *l'Hôpital's rule* (to handle indeterminate forms)
- *squeeze theorem*

Another useful tool is to examine the behavior of aptly chosen subsequences of a given sequence $\{a_n\}_{n=1}^{\infty}$. Specifically, the following two criteria are useful:

1. If the subsequence of odd-indexed terms and the subsequence of even-indexed terms both converge to the same limit L, then $\lim_{n \to \infty} a_n = L$.

2. If you can form two subsequences of $\{a_n\}_{n=1}^{\infty}$ that converge to *different* values, then the original sequence $\{a_n\}_{n=1}^{\infty}$ diverges.

Definition of Infinite Series

- Let $\{a_k\}$ be a sequence of real numbers. Now, define a new sequence of *partial sums* $\{S_n\}$ by $S_n = \sum_{k=1}^{n} a_k$. Then the pair $\left(\{a_n\}, \{S_n\} \right)$ is called an *infinite series*. It is customary to write $\sum_{k=1}^{\infty} a_k$ to represent such an infinite series.

- An infinite series is *convergent* to $L(< \infty)$ if there exists $\lim_{n \to \infty} S_n$. In such case, we write $\sum_{k=1}^{\infty} a_k = L$.

- If either $\lim_{n \to \infty} S_n$ does not exist or $\lim_{n \to \infty} S_n = \pm\infty$, then we say the infinite series is *divergent*.

A natural question to ask is, "When does a series converge?"

Geometric Series

A *geometric series* is one of the form $\sum_{k=1}^{\infty} ar^{k-1}$, where $a \neq 0$. Such a series converges (with sum $\frac{a}{1-r}$) if and only if $|r| < 1$.

Remark: What about a series of the form $\sum_{k=m+1}^{\infty} ar^{k-1}$? It is essentially geometric, regardless of the fact that the first term is not simply a. We can argue in a manner similar to the preceding to arrive at the same necessary and sufficient condition for convergence, but the sum will be slightly different. Indeed, for $|r| < 1$, we have

$$\frac{a}{1-r} = \sum_{k=1}^{\infty} ar^{k-1} = \sum_{k=1}^{m} ar^{k-1} + \sum_{k=m+1}^{\infty} ar^{k-1}$$

so that

$$\sum_{k=m+1}^{\infty} ar^{k-1} = \frac{a}{1-r} - \underbrace{\sum_{k=1}^{m} ar^{k-1}}_{= \frac{a-ar^m}{1-r}} = \frac{a}{1-r} - \left[\frac{a}{1-r} - \frac{ar^m}{1-r} \right] = \frac{ar^m}{1-r}$$

(Notice that the sum itself is of the form $\frac{\text{First term of the series}}{1-r}$.)

p-Series

The series $\sum_{n=1}^{\infty} \frac{1}{n^p}$ converges if and only if $p > 1$. (If $p = 1$, the series is called the *harmonic series*.)

Linearity of Convergent Series

Suppose that $\sum_{k=1}^{\infty} a_k$ and $\sum_{k=1}^{\infty} b_k$ *both converge* and let $c \in \mathbb{R}$. Then,

- $\sum_{k=1}^{\infty} ca_k$ converges and $= c \sum_{k=1}^{\infty} a_k$.

- $\sum_{k=1}^{\infty} (a_k \pm b_k)$ converges and $= \sum_{k=1}^{\infty} a_k \pm \sum_{k=1}^{\infty} b_k$.

The *n*th Term Test for Divergence

If $\lim_{k \to \infty} a_k \neq 0$, then $\sum_{k=1}^{\infty} a_k$ diverges.

Ratio Test

Consider the series $\sum_{k=1}^{\infty} a_k$ and let $\lim_{k \to \infty} \left| \frac{a_{k+1}}{a_k} \right| = \rho$.

- If $\rho < 1$, then $\sum_{k=1}^{\infty} a_k$ converges (absolutely).

- If $\rho > 1$, then $\sum_{k=1}^{\infty} a_k$ diverges.

- If $\rho = 1$, then the test is inconclusive.

Limit Comparison Test

Suppose that $\sum\limits_{k=1}^{\infty} a_k$ and $\sum\limits_{k=1}^{\infty} b_k$ are given series. If $\lim\limits_{k \to \infty} \frac{a_k}{b_k} > 0$, then $\sum\limits_{k=1}^{\infty} a_k$ converges (or diverges) if and only if $\sum\limits_{k=1}^{\infty} b_k$ converges (or diverges).

Ordinary Comparison Test

Suppose that $\sum\limits_{k=1}^{\infty} a_k$ and $\sum\limits_{k=1}^{\infty} b_k$ are given series and that $a_k \le b_k$ for all values of k larger than some integer N. Then,

- If $\sum\limits_{k=1}^{\infty} b_k$ converges, then $\sum\limits_{k=1}^{\infty} a_k$ converges.

- If $\sum\limits_{k=1}^{\infty} a_k$ diverges, then $\sum\limits_{k=1}^{\infty} b_k$ diverges.

The terms used to define an infinite series $\sum\limits_{k=1}^{\infty} a_k$ can be positive or negative. There are two levels of convergence, namely *conditional convergence* and *absolute convergence*, defined as follows:

- The series $\sum\limits_{k=1}^{\infty} a_k$ is absolutely convergent if $\sum\limits_{k=1}^{\infty} |a_k|$ converges.

- The series $\sum\limits_{k=1}^{\infty} a_k$ is conditionally convergent if $\sum\limits_{k=1}^{\infty} |a_k|$ diverges but $\sum\limits_{k=1}^{\infty} a_k$ converges.

Of particular utility if the *alternating series test*.

Alternating Series Test

Consider the series $\sum\limits_{k=1}^{\infty} (-1)^k a_k$, where $a_k \ge 0$, for all k. If the sequence $\{a_k\}$ decreases to zero, then $\sum\limits_{k=1}^{\infty} (-1)^k a_k$ converges.

Questions

396. Carefully explain whether the sequence $\left\{ \dfrac{\arctan{(n)}}{e^{3n}} \right\}_{n=1}^{\infty}$ converges or diverges. If it is convergent, find its limit.

397. Let $a_n = \dfrac{\sin\left(\frac{n\pi}{2}\right)}{2n+1}$, $n = 1, 2, 3, \ldots$; which of the following characterizations of the sequence $\left\{ a_n \right\}_{n=1}^{\infty}$ is correct?

a. $\left\{ a_n \right\}_{n=1}^{\infty}$ is bounded and monotone.

b. $\left\{ a_n \right\}_{n=1}^{\infty}$ is bounded and convergent.

c. $\left\{ a_n \right\}_{n=1}^{\infty}$ is monotone and convergent.

d. $\left\{ a_n \right\}_{n=1}^{\infty}$ is bounded, monotone, and convergent.

398. Determine whether the series $\displaystyle\sum_{k=2}^{\infty} \left[4\left(\frac{1}{3}\right)^k - 5 \cdot \frac{2^{2k}}{7^{3k-1}} \right]$ converges or diverges.

399. Find the sum of the series:

$$\frac{2}{7} - \left(\frac{2}{7}\right)^2 + \left(\frac{2}{7}\right)^3 - \left(\frac{2}{7}\right)^4 + \left(\frac{2}{7}\right)^5 - \left(\frac{2}{7}\right)^6 + - \ldots$$

400. Express the decimal $3.125125125125\ldots$ as a fraction.

401. Determine if the series $1 + \dfrac{1}{\sqrt{2}} + \dfrac{1}{\sqrt{3}} + \dfrac{1}{\sqrt{4}} + \ldots$ converges or diverges.

402. Determine if the sequence of partial sums of the following series converges or diverges. Then, use this to make a conclusion about the infinite series.

$$1 - \frac{1}{2} + \frac{1}{2} - \frac{1}{4} + \frac{1}{4} - \frac{1}{8} + \frac{1}{8} - \frac{1}{16} + \frac{1}{16} - + \ldots$$

403. True or false? The series $\displaystyle\sum_{n=1}^{\infty} (-1)^{n+1} \frac{n^2}{e^n}$ converges absolutely.

404. Determine if the series $\displaystyle\sum_{n=3}^{\infty} \frac{4n+3}{\left(5n^2-2\right)^2}$ converges or diverges.

405. Determine if the series $\displaystyle\sum_{n=2}^{\infty} 2^{\frac{1}{n}}$ converges or diverges.

406. True or false? If $\lim\limits_{k \to \infty} a_k = 0$, then $\sum\limits_{k=1}^{\infty} a_k$ converges.

407. True or false? If $\lim\limits_{k \to \infty} a_k = 0$, then $\{a_k\}$ converges.

408. True or false? If $\sum\limits_{k=1}^{\infty} a_k$ diverges, then $\{a_k\}$ does not have a limit.

409. True or false? If $\{a_k\}$ converges, then $\{a_k\}$ is bounded.

410. True or false? Suppose that $0 \le a_k \le b_k$, for all k. If $\sum\limits_{k=1}^{\infty} a_k$ converges, then $\sum\limits_{k=1}^{\infty} b_k$ must also converge.

411. True or false? Assume that $a_k \ge 0$ for each k, and let $S_n = \sum\limits_{k=1}^{n} a_k$. If $\lim\limits_{n \to \infty} S_n = 0$, then $\sum\limits_{k=1}^{\infty} a_k$ converges.

412. True or false? $\sum\limits_{n=1}^{\infty}\left[\dfrac{1}{n\sqrt{n}} - 7\left(-\dfrac{2}{3}\right)^{n+1} \right]$ is a convergent series.

413. True or false? Assume that $a_k \ge 0$ for each k. If $\lim\limits_{k \to \infty} a_k = 0$, then $\lim\limits_{k \to \infty} (-1)^k a_k = 0$.

414. The 35th term of the sequence $\frac{1}{2}, -\frac{1}{4}, \frac{1}{8}, \ldots$ is which of the following?
   a. $\dfrac{1}{2^{35}}$
   b. $-\dfrac{1}{2^{35}}$
   c. $\dfrac{1}{2^{36}}$
   d. $-\dfrac{1}{2^{36}}$

415. True or false? The sequence whose nth term is defined by $a_n = \dfrac{\cos n}{5^n}$, $n \ge 1$, converges.

416. True or false? The sequence $\{a_n\}_{n=1}^{\infty}$ whose nth term is defined recursively by
$$\begin{cases} a_0 = 9 \\ a_n = \frac{1}{3} a_{n-1}, \ n \ge 1 \end{cases}$$
is divergent.

417. The sequence $\{a_n\}_{n=1}^{\infty}$ whose nth term is defined by $a_n = \frac{3n+2}{1-6n}, n \geq 1$, converges to which of the following limits?

a. 0

b. $-\frac{1}{2}$

c. 2

d. 3

418. True or false? The sequence $\{a_n\}_{n=1}^{\infty}$ whose nth term is defined by

$a_n = \frac{(-1)^n n}{n+1}, n \geq 1$ is divergent.

419. Show that the sequence $\{a_n\}_{n=1}^{\infty}$ whose nth term is defined by

$a_n = \sum_{k=1}^{n} \left[\frac{1}{k+1} - \frac{1}{k} \right], n \geq 1$, converges, and find its limit.

420. Which of the following is an accurate characterization of the sequence $\{a_n\}_{n=1}^{\infty}$ whose nth term is defined by $a_n = 2 + \frac{1-\sqrt{n}}{1-n}$?

a. $\{a_n\}_{n=1}^{\infty}$ converges to 1.

b. $\{a_n\}_{n=1}^{\infty}$ converges to 2.

c. $\{a_n\}_{n=1}^{\infty}$ converges to 3.

d. $\{a_n\}_{n=1}^{\infty}$ diverges.

421. If $a_n = 2 + (-1)^n$, $n \geq 1$, then which of the following is an accurate characterization of the series $\sum\limits_{n=1}^{\infty} a_n$?

a. The series $\sum\limits_{n=1}^{\infty} a_n$ converges with a sum of 2.

b. The series $\sum\limits_{n=1}^{\infty} a_n$ converges with a sum of 3.

c. The series $\sum\limits_{n=1}^{\infty} a_n$ diverges.

d. None of the above is accurate.

422. Which of the following is an accurate characterization of the series
$\sum_{n=2}^{\infty} \frac{3^{3n+1}}{27^{2n-1}}$?

   a. The series is convergent with sum $\frac{27}{26}$.

   b. The series is convergent with sum $\frac{3}{26}$.

   c. The series is convergent with sum $\frac{81}{26}$.

   d. The series is divergent.

423. Which of the following is an accurate characterization of the series
$\sum_{n=1}^{\infty} \left[1 + \left(\frac{2}{3} \right)^{n-1} \right]$?

   a. The series is convergent with sum 4.

   b. The series is convergent with sum 3.

   c. The series is convergent with sum 2.

   d. The series is divergent.

424. True or false? It follows from the limit comparison test that the series
$\sum_{n=1}^{\infty} \frac{n}{3n^2 + 1}$ diverges.

425. True or false? It follows from the limit comparison test that the series
$\sum_{n=1}^{\infty} \frac{n+1}{4n^3 + n + 1}$ diverges.

426. Use the ordinary comparison test to show that the series $\sum_{n=2}^{\infty} e^{-n} \sin^2(n)$
converges.

427. Use the ordinary comparison test to show that the series $\sum_{n=1}^{\infty} \frac{1 + e^{-2n}}{n+1}$
diverges.

428. Use the ratio rest to determine if the series $\sum_{n=1}^{\infty} \frac{2^n n^3}{n!}$ converges or
diverges.

429. Use the ratio rest to determine if the series $\sum_{n=1}^{\infty} \frac{3^n n}{(2n)!}$ converges or
diverges.

430. True or false? The series $\sum_{n=1}^{\infty} \frac{(-1)^n}{3n}$ conditionally converges.

431. True or false? The series $\sum_{n=1}^{\infty} [\ln(en) - \ln(n+2)]$ converges.

432. True or false? If $\lim_{n \to \infty} a_n = 1$, then the series $\sum_{n=1}^{\infty} \dfrac{2a_n}{1 + 5a_n}$ converges.

433. Which of the following is an accurate characterization of the series $\sum_{n=0}^{\infty} -2\left(\dfrac{5}{6}\right)^{n+1}$?
   a. The series converges with a sum of −10.
   b. The series converges with a sum of −12.
   c. The series converges with a sum of 6.
   d. The series diverges.

434. Determine if the following sequence converges or diverges:
   $-2,\ 1,\ -1,\ \frac{1}{2},\ -2,\ \frac{1}{4},\ -1, \ldots$

435. True or false? The sequence whose nth term is given by $a_n = n \sin\left(\frac{1}{n}\right)$, $n \geq 1$, converges.

Answers

396. First, since the graph of $y = \arctan(x)$ is bounded between the horizontal lines $y = \frac{\pi}{2}$ and $y = -\frac{\pi}{2}$ (which play the role of the horizontal asymptotes for the graph), it follows that $-\frac{\pi}{2} \leq \arctan n \leq \frac{\pi}{2}$, for all integers n. Moreover, since $e^{3n} > 0$ for all integers n, we can divide all parts of the previous inequality by e^{3n} to obtain:

$\frac{-\frac{\pi}{2}}{e^{3n}} \leq \frac{\arctan n}{e^{3n}} \leq \frac{\frac{\pi}{2}}{e^{3n}}$, for all integers n

Now, since $\lim_{n \to \infty} e^{3n} = \infty$, it follows that $\lim_{n \to \infty} \left(\frac{-\frac{\pi}{2}}{e^{3n}} \right) = \lim_{n \to \infty} \left(\frac{\frac{\pi}{2}}{e^{3n}} \right) = 0$ (since the top of the fraction is constant and the denominator becomes arbitrarily large as n goes to infinity). Thus, we conclude by the squeeze theorem that $\left\{ \frac{\arctan n}{e^{3n}} \right\}$ converges to the limit 0.

397. b. First, note that the first few values of $\sin\left(\frac{n\pi}{2}\right)$, for $n = 1, 2, 3, \ldots$ are given by:

$1, 0, -1, 0, 1, 0, -1, \ldots$

Hence, the first few terms of the sequence $\{a_n\}$ are as follows:

$\frac{1}{2(1)+1}, 0, \frac{-1}{2(3)+1}, 0, \frac{1}{2(5)+1}, 0, \frac{-1}{2(7)+1}, \ldots$

The terms of this sequence are cyclic with the pattern being that the terms start at a positive value, then move to zero, then to a negative value, then to zero, and then back to a positive value. This pattern repeats ad infinitum. Hence, the sequence cannot be monotonic. However, the terms *do* get closer to zero from both the negative and positive directions as n goes to infinity. As such, the sequence is bounded and, in fact, convergent to zero.

398. It would be nice to be able to consider the given series as a sum of two series and investigate each one separately. Indeed, we must determine if we can use linearity. We investigate each of the two series independently and then draw our conclusions using the preceding remarks.

First, observe that $\sum\limits_{k=2}^{\infty} 4\left(\frac{1}{3}\right)^k$ is a convergent geometric series with

sum $\dfrac{4\left(\frac{1}{3}\right)^2}{1-\frac{1}{3}} = \dfrac{2}{3}.$

Also, it is not immediately obvious that $\sum\limits_{k=2}^{\infty} -5\cdot\dfrac{2^{2k}}{7^{3k-1}}$ is geometric, but after we perform some elementary algebraic computations, we shall determine that it is indeed geometric. Indeed, observe that

$$\underbrace{\sum_{k=2}^{\infty} -5\cdot\frac{2^{2k}}{7^{3k-1}}}_{\text{needs to be of the form }\sum\limits_{k} ar^k} = \sum_{k=2}^{\infty} -5\cdot\frac{\left(2^2\right)^k}{\left(7^3\right)^k\cdot 7^{-1}} = \sum_{k=2}^{\infty} -35\left(\frac{4}{343}\right)^k$$

and this *is* a convergent geometric series with sum $-\dfrac{35\left(\frac{4}{343}\right)^2}{1-\frac{4}{343}}.$

Consequently, we can use the linearity result to conclude that the original series $\sum\limits_{k=2}^{\infty}\left[4\left(\frac{1}{3}\right)^k - 5\cdot\frac{2^{2k}}{7^{3k-1}}\right]$ converges and has sum

$\dfrac{2}{3} - \dfrac{35\left(\frac{4}{343}\right)^2}{1-\frac{4}{343}}.$

399. The series $\frac{2}{7} - \left(\frac{2}{7}\right)^2 + \left(\frac{2}{7}\right)^3 - \left(\frac{2}{7}\right)^4 + \left(\frac{2}{7}\right)^5 - \left(\frac{2}{7}\right)^6 + - \cdots$ can be rewritten as a constant times a convergent geometric series, as follows:

$$\sum_{n=0}^{\infty} (-1)^n \left(\frac{2}{7}\right)^{n+1} = \frac{2}{7} \sum_{n=0}^{\infty} (-1)^n \left(\frac{2}{7}\right)^n = \frac{2}{7} \sum_{n=0}^{\infty} \left(-\frac{2}{7}\right)^n = \frac{2}{7} \cdot \frac{1}{1 - \left(-\frac{2}{7}\right)} = \frac{2}{7} \cdot \frac{7}{9} = \frac{2}{9}$$

Thus, the series converges with sum $\frac{2}{9}$.

400. We write this decimal as a convergent geometric series. To this end, observe that

$$3.125125125125\ldots = 3 + 0.125 + 0.000125 + 0.000000125$$
$$+ \ 0.000000000125 + \ldots$$
$$= 3 + \frac{125}{10^3} + \frac{125}{10^6} + \frac{125}{10^9} + \frac{125}{10^{12}} + \ldots$$
$$= 3 + \sum_{n=1}^{\infty} 125 \cdot \frac{1}{\left(10^3\right)^n}$$
$$= 3 + \sum_{n=1}^{\infty} 125 \cdot \left(\frac{1}{1,000}\right)^n$$
$$= 3 + \frac{125 \cdot \left(\frac{1}{1,000}\right)}{1 - \frac{1}{1,000}}$$
$$= 3 + \frac{125}{999}$$
$$= \frac{3,122}{999}$$

401. Note that the series $1 + \frac{1}{\sqrt{2}} + \frac{1}{\sqrt{3}} + \frac{1}{\sqrt{4}} + \cdots$ can be written in the more compact form $\sum_{n=1}^{\infty} \frac{1}{\sqrt{n}}$. Moreover, observe that $\frac{1}{n} \leq \frac{1}{\sqrt{n}}$ for all positive integers n; this is true because the denominator of the fraction on the left side of the inequality is larger than the denominator of the fraction on the right side, and the two fractions have the same numerator. It is known that the harmonic series $\sum_{n=1}^{\infty} \frac{1}{n}$ diverges. Hence, we conclude from the ordinary comparison test that $\sum_{n=1}^{\infty} \frac{1}{\sqrt{n}}$ diverges.

402. Let S_n = sum of the first n terms of the given series.

The first few terms of the sequence of partial sums $\{S_n\}$ are computed as follows:

$S_1 = 1$

$S_2 = 1 - \frac{1}{2} = \frac{1}{2}$

$S_3 = 1 - \frac{1}{2} + \frac{1}{2} = 1$

$S_4 = 1 - \frac{1}{2} + \frac{1}{2} - \frac{1}{4} = \frac{3}{4}$

$S_5 = 1 - \frac{1}{2} + \frac{1}{2} - \frac{1}{4} + \frac{1}{4} = 1$

$S_6 = 1 - \frac{1}{2} + \frac{1}{2} - \frac{1}{4} + \frac{1}{4} - \frac{1}{8} = \frac{7}{8}$

$S_7 = 1 - \frac{1}{2} + \frac{1}{2} - \frac{1}{4} + \frac{1}{4} - \frac{1}{8} + \frac{1}{8} = 1$

From the pattern that emerges, we see that the following is an explicit formula for the nth term of $\{S_n\}$:

$$S_n = \begin{cases} 1, & \text{if } n \text{ is odd} \\ \dfrac{2^{\frac{n}{2}} - 1}{2^{\frac{n}{2}}}, & \text{if } n \text{ is even} \end{cases}$$

Since the subsequence of odd-indexed terms and the subsequence of even-indexed terms both approach 1 as n goes to infinity, we conclude that $\lim_{n \to \infty} S_n = 1$. Hence, by definition of convergence of an infinite series, we conclude that $\sum_{n=1}^{\infty} a_n$ converges with sum 1.

403. True. The series $\sum_{n=1}^{\infty} (-1)^{n+1} \dfrac{n^2}{e^n}$ converges absolutely because using the ratio test yields

$$\lim_{n \to \infty} \left| \frac{(-1)^{n+2} \frac{(n+1)^2}{e^{n+1}}}{(-1)^{n+1} \frac{n^2}{e^n}} \right| = \lim_{n \to \infty} \frac{(n+1)^2}{e^{n+1}} \cdot \frac{e^n}{n^2} = \lim_{n \to \infty} \frac{(n+1)^2}{n^2} \cdot \frac{1}{e} = \frac{1}{e} < 1$$

404. First, note that the series $\sum_{n=3}^{\infty} \frac{1}{n^3}$ is a convergent p-series (since $p = 3 > 1$). Moreover, observe that

$$\lim_{n\to\infty}\frac{\frac{4n+3}{(5n^2-2)^2}}{\frac{1}{n^3}} = \lim_{n\to\infty}\frac{(4n+3)n^3}{(5n^2-2)^2} = \lim_{n\to\infty}\frac{4n^4+3n^3}{25n^4-20n^2+4} = \frac{4}{25} > 0$$

Thus, the limit comparison test implies that the series $\sum_{n=3}^{\infty}\frac{4n+3}{(5n^2-2)^2}$ converges.

405. Note that the limit of the nth term of the series is $\lim_{n\to\infty} 2^{\frac{1}{n}} = 2^0 = 1$. Since this is not zero, we conclude immediately from the nth term test that the series $\sum_{n=2}^{\infty} 2^{\frac{1}{n}}$ diverges.

406. False. The harmonic series $\sum_{k=1}^{\infty}\frac{1}{k}$ is a counterexample to the statement since $\lim_{k\to\infty}\frac{1}{k} = 0$, yet $\sum_{k=1}^{\infty}\frac{1}{k}$ is known to diverge.

407. True. This is true by definition of convergence. If a sequence has a limit of zero, then it converges to zero. (Compare this to Question 406. The two statements are very different because the convergences of different quantities are considered in the two statements.)

408. False. The harmonic series $\sum_{k=1}^{\infty}\frac{1}{k}$ is a counterexample to the statement since $\sum_{k=1}^{\infty}\frac{1}{k}$ is known to diverge, yet $\lim_{k\to\infty}\frac{1}{k} = 0$, so that $\left\{\frac{1}{k}\right\}_{k=1}^{\infty}$ converges.

409. True. A formal proof of this statement is beyond the scope of this book, but the essence of the argument is as follows. If a sequence $\{a_k\}$ converges to L, then the terms must be very close to the limit value L for all k from some point on. Precisely, the terms a_k must lie in an interval centered at L, for all $k > N$. The only terms of the original sequence that might not lie in this interval are those whose index is one of $1, 2, \ldots,$ $N - 1$. Since there are only finitely many of these values, one of the terms of the sequence with such an index is larger than all of the others in absolute value. Thus, these terms are also contained within an interval centered at 0 with radius equal to this extreme value. Consequently, we can conclude that the terms don't march off toward infinity, but rather remain bounded as k goes to infinity.

Alternatively, the contrapositive of the statement is: "If $\{a_k\}$ is not bounded, then $\{a_k\}$ diverges." This statement is equivalent to the given one, and is arguably somewhat easier to think about. Indeed, if the terms of the sequence are unbounded, then as the index k goes to infinity, the terms of at least a subsequence go to plus or minus infinity. Thus, we can never determine a value L to which *all* of the terms from some point on are very close to L.

410. False. Bounding a given series *from below* by one that *converges* does nothing to control the behavior of the given series. For instance, the p-series $\sum_{n=1}^{\infty} \frac{1}{n^2}$ converges (since $p = 2 > 1$) and clearly, the series $\sum_{n=1}^{\infty} n$ diverges (by the nth term test). But, note that $0 \leq \frac{1}{n^2} \leq n$ for all positive integers n. Therefore, the given statement is false in a big way!

411. True. This follows immediately from the definition of convergence of an infinite series. Here, it is given that the sequence of partial sums $S_n = \sum_{k=1}^{n} a_k$ converges to zero. Hence, by definition, the series $\sum_{k=1}^{\infty} a_k$ converges (to zero).

412. True. Observe that the series $\sum_{n=1}^{\infty}\frac{1}{n\sqrt{n}}=\sum_{n=1}^{\infty}\frac{1}{n^{\frac{3}{2}}}$ is a convergent p-series

(since $p=\frac{3}{2}>1$). Also, the series $\sum_{n=1}^{\infty}7\left(-\frac{2}{3}\right)^{n+1}$ is a convergent

geometric series. As such, by linearity, we conclude that

$\sum_{n=1}^{\infty}\left[\frac{1}{n\sqrt{n}}-7\left(-\frac{2}{3}\right)^{n+1}\right]$ is a convergent series. (Note that we cannot

find the sum of this series because there is no known formula for the

sum of a convergent p-series.)

413. True. Observe that $-a_k\leq(-1)^k a_k\leq a_k$ for any k. Also, since $\lim_{k\to\infty}a_k=0$,

it follows from linearity of limits that $\lim_{k\to\infty}-a_k=-\lim_{k\to\infty}a_k=0$. Thus,

we conclude from the squeeze theorem that $\lim_{k\to\infty}(-1)^k a_k=0$.

414. a. This sequence can be expressed in the more compact form

$a_n=\frac{(-1)^{n+1}}{2^n}$, $n\geq1$. Hence, the 35th term is $a_{35}=\frac{(-1)^{35+1}}{2^{35}}=\frac{1}{2^{35}}$.

415. True. Note that $-1\leq\cos n\leq1$ for all integers n. Also, since $5^n>0$, for all

integers n, we can divide all parts of the previous inequality by 5^n to

obtain:

$-\frac{1}{5^n}\leq\frac{\cos n}{5^n}\leq\frac{1}{5^n}$

Since $\lim_{n\to\infty}-\frac{1}{5^n}=\lim_{n\to\infty}\frac{1}{5^n}=0$, we conclude from the squeeze theorem that

the sequence whose nth term is defined by $a_n=\frac{\cos n}{5^n}$, $n\geq1$, converges to

zero.

416. False. In words, the terms of the given sequence are formed by dividing

the previous term in the sequence by 3. The first few terms of the

sequence are as follows:

$9,\ \frac{9}{3},\ \frac{9}{3^2},\ \frac{9}{3^3},\ \frac{9}{3^4},\ \ldots$

We can therefore express the recursively defined sequence as one for

which the nth term is specified by $a_n=\frac{9}{3^n}=\frac{3^2}{3^n}=\frac{1}{3^{n-2}}$, $n\geq0$. The

limit of this sequence is zero. Hence, it is convergent.

417. b. $\lim_{n\to\infty}\frac{3n+2}{1-6n}=\lim_{n\to\infty}\frac{3n+2}{1-6n}\cdot\frac{\frac{1}{n}}{\frac{1}{n}}=\lim_{n\to\infty}\frac{3+\frac{2}{n}}{\frac{1}{n}-6}=\frac{3}{-6}=-\frac{1}{2}$

418. True. The subsequence consisting of the odd-indexed terms of the sequence are all negative and are described by the explicit formula $-\frac{n}{n+1}$; this subsequence converges to -1. Similarly, the subsequence consisting of the even-indexed terms of the sequence are all positive and are described by the explicit formula $\frac{n}{n+1}$; this subsequence converges to 1. Since these two subsequences do not converge to the same limit value, we conclude that the original sequence $\left\{a_n\right\}_{n=1}^{\infty}$ must diverge.

419. It is particularly helpful in this case to write out the first few terms in an attempt to obtain a more simplified expression for the nth term. To this end, we have:

$$a_1 = \left(\tfrac{1}{2} - 1\right)$$

$$a_2 = \left(\tfrac{1}{\cancel{2}} - 1\right) + \left(\tfrac{1}{3} - \tfrac{1}{\cancel{2}}\right) = \tfrac{1}{3} - 1$$

$$a_3 = \left(\tfrac{1}{\cancel{2}} - 1\right) + \left(\tfrac{1}{\cancel{3}} - \tfrac{1}{\cancel{2}}\right) + \left(\tfrac{1}{4} - \tfrac{1}{\cancel{3}}\right) = \tfrac{1}{4} - 1$$

The pattern that emerges enables us to express the sequence $a_n = \sum_{k=1}^{n}\left[\frac{1}{k+1} - \frac{1}{k}\right], n \geq 1$, in the simplified form $a_n = \frac{1}{n+1} - 1$, and it is not difficult to see that $\lim\limits_{n\to\infty} a_n = \lim\limits_{n\to\infty}\left(\frac{1}{n+1} - 1\right) = -1$. This shows that the sequence converges.

420. b. The strategy is to rationalize the numerator of the fractional portion of the expression for the nth term, and then cancel a factor common to the numerator and denominator, as follows:

$$\lim\limits_{n\to\infty} a_n = \lim\limits_{n\to\infty}\left(2 + \frac{1 - \sqrt{n}}{1 - n}\right) = \lim\limits_{n\to\infty}\left(2 + \frac{1 - \sqrt{n}}{1 - n} \cdot \frac{1 + \sqrt{n}}{1 + \sqrt{n}}\right) = \lim\limits_{n\to\infty}\left(2 + \frac{\cancel{1 - n}}{\cancel{(1 - n)}\left(1 + \sqrt{n}\right)}\right)$$

$$= \lim\limits_{n\to\infty}\left(2 + \frac{1}{1+\sqrt{n}}\right) = 2 + 0 = 2$$

421. c. Note that the terms of the sequence $\left\{a_n\right\}$ can be expressed by the list $1, 3, 1, 3, \ldots$ Thus, the sequence does not converge. Specifically, $\lim\limits_{n\to\infty} a_n \neq 0$. As a result, we conclude from the nth term test for divergence that the series $\sum_{n=1}^{\infty} a_n$ diverges.

422. **b.** We rewrite the series $\sum\limits_{n=2}^{\infty} \frac{3^{3n+1}}{27^{2n-1}}$ in the form of a convergent

geometric series using the exponent rules, as follows:

$$\sum_{n=2}^{\infty} \frac{3^{3n+1}}{27^{2n-1}} = \sum_{n=2}^{\infty} \frac{3^{3n} \cdot 3}{\left(3^3\right)^{2n-1}} = \sum_{n=2}^{\infty} \frac{3^{3n} \cdot 3}{\left(3^3\right)^{2n} \left(3^3\right)^{-1}} = \sum_{n=2}^{\infty} \frac{3^{3n} \cdot 3^4}{3^{6n}} = \sum_{n=2}^{\infty} 81 \cdot 3^{-3n}$$

$$= \sum_{n=2}^{\infty} 81 \cdot \left(\frac{1}{27}\right)^n$$

The sum of this convergent series is given by:

$$\frac{81 \cdot \left(\frac{1}{27}\right)^2}{1 - \frac{1}{27}} = \frac{81 \cdot \left(\frac{1}{27}\right)^2}{\frac{26}{27}} = \frac{\frac{1}{9}}{\frac{26}{27}} = \frac{1}{9} \cdot \frac{27}{26} = \frac{3}{26}$$

423. **d.** Note that $\lim\limits_{n\to\infty} \left[1 + \left(\frac{2}{3}\right)^{n-1} \right] = 1 + 0 = 1$. Since this value is not zero, we

conclude immediately from the nth term test for divergence that the

series $\sum\limits_{n=1}^{\infty} \left[1 + \left(\frac{2}{3}\right)^{n-1} \right]$ diverges.

424. True. First, note that the series $\sum\limits_{n=1}^{\infty} \frac{1}{n}$ is the divergent harmonic series.

Moreover, observe that

$$\lim_{n\to\infty} \frac{\frac{n}{3n^2+1}}{\frac{1}{n}} = \lim_{n\to\infty} \frac{n^2}{3n^2+1} = \lim_{n\to\infty} \frac{n^2}{3n^2+1} \cdot \frac{\frac{1}{n^2}}{\frac{1}{n^2}} = \lim_{n\to\infty} \frac{1}{3 + \frac{1}{n^2}} = \frac{1}{3} > 0$$

Thus, the limit comparison test implies that the series $\sum\limits_{n=1}^{\infty} \frac{n}{3n^2+1}$
diverges.

425. False. First, note that the series $\sum\limits_{n=1}^{\infty} \frac{1}{n^2}$ is a convergent p-series

(since $p = 2 > 1$). Moreover, observe that

$$\lim_{n\to\infty} \frac{\frac{n+1}{4n^3+n+1}}{\frac{1}{n^2}} = \lim_{n\to\infty} \frac{n^3+n^2}{4n^3+n+1} = \lim_{n\to\infty} \frac{n^3+n^2}{4n^3+n+1} \cdot \frac{\frac{1}{n^3}}{\frac{1}{n^3}} = \lim_{n\to\infty} \frac{1+\frac{1}{n}}{4+\frac{1}{n^2}+\frac{1}{n^3}} = \frac{1}{4} > 0$$

Thus, the limit comparison test implies that the series $\sum\limits_{n=1}^{\infty} \frac{n+1}{4n^3+n+1}$
converges.

426. First, note that $e^{-n} \sin^2 n \geq 0$, for all integers n. Also, $\sin^2 n \leq 1$ for all integers n. Thus, $e^{-n} \sin^2 n \leq e^{-n}$, for all integers n, so the series also obeys the same inequality, namely $\sum_{n=2}^{\infty} e^{-n} \sin^2(n) \leq \sum_{n=2}^{\infty} e^{-n}$. Now, note that $\sum_{n=2}^{\infty} e^{-n} = \sum_{n=2}^{\infty} \left(\frac{1}{e}\right)^n$ is a convergent geometric series. Thus, we conclude from the ordinary comparison test that the series $\sum_{n=2}^{\infty} e^{-n} \sin^2(n)$ converges.

427. First, note that $\frac{1+e^{-2n}}{n+1} > 0$ for all positive integers n. Also, since $\frac{e^{-2n}}{n+1} > 0$ for all positive integers n, it follows that $\frac{1}{n+1} \leq \frac{1+e^{-2n}}{n+1}$, for all positive integers n. But, $\sum_{n=1}^{\infty} \frac{1}{n+1}$ is one term shy of a divergent harmonic series. Since subtracting a single term from a divergent series does not render the resulting series convergent, we conclude that $\sum_{n=1}^{\infty} \frac{1}{n+1}$ is divergent. Thus, we conclude from the ordinary comparison test that the series $\sum_{n=1}^{\infty} \frac{1+e^{-2n}}{n+1}$ diverges.

428. The series $\sum_{n=1}^{\infty} \frac{2^n n^3}{n!}$ converges absolutely because using the ratio test yields

$$\lim_{n \to \infty} \left| \frac{\frac{2^{n+1}(n+1)^3}{(n+1)!}}{\frac{2^n n^3}{n!}} \right| = \lim_{n \to \infty} \frac{2^{n+1}(n+1)^3}{(n+1)!} \cdot \frac{n!}{2^n n^3} = \lim_{n \to \infty} \frac{2^n \cdot 2 \cdot (n+1)^3}{(n+1) \cdot n!} \cdot \frac{n!}{2^n n^3}$$

$$= \lim_{n \to \infty} \frac{2(n+1)^2}{n^3} = 0 < 1$$

429. The series $\sum_{n=1}^{\infty} \frac{3^n n}{(2n)!}$ converges absolutely because using the ratio test yields

$$\lim_{n \to \infty} \left| \frac{\frac{3^{n+1}(n+1)}{(2(n+1))!}}{\frac{3^n n}{(2n)!}} \right| = \lim_{n \to \infty} \frac{3^{n+1}(n+1)}{(2(n+1))!} \cdot \frac{(2n)!}{3^n n} = \lim_{n \to \infty} \frac{3^n \cdot 3 \cdot (n+1)}{(2n+2)!} \cdot \frac{(2n)!}{3^n n}$$

$$= \lim_{n \to \infty} \frac{3^n \cdot 3 \cdot (n+1)}{(2n+2)(2n+1)(2n)!} \cdot \frac{(2n)!}{3^n n} = \lim_{n \to \infty} \frac{3(n+1)}{2(n+1)(2n+1)n}$$

$$= \lim_{n \to \infty} \frac{3}{2(2n+1)n} = 0 < 1$$

430. True. We must show that $\sum_{n=1}^{\infty} \frac{(-1)^n}{3n}$ converges (using the alternating

series test), and that $\sum_{n=1}^{\infty} \frac{1}{3n}$ diverges. To begin, note that the sequence

$a_n = \frac{1}{3n}$, $n \geq 1$, is comprised of only positive terms and is a decreasing
sequence heading toward 0. We therefore conclude from the

alternating series test that $\sum_{n=1}^{\infty} \frac{(-1)^n}{3n}$ converges. Next, since $\sum_{n=1}^{\infty} \frac{1}{n}$ is the

divergent harmonic series and $\lim_{n \to \infty} \frac{\frac{1}{3n}}{\frac{1}{n}} = \lim_{n \to \infty} \frac{1}{3n} \cdot n = \frac{1}{3} > 0$, we conclude

from the limit comparison test that the series $\sum_{n=1}^{\infty} \left| \frac{(-1)^n}{3n} \right| = \sum_{n=1}^{\infty} \frac{1}{3n}$ diverges.

Thus, we conclude that the given series $\sum_{n=1}^{\infty} \frac{(-1)^n}{3n}$ is conditionally
convergent.

431. False. Using the logarithm rules, observe that

$\ln(en) - \ln(n + 2) = \ln\left(\frac{en}{n + 2} \right)$. Moreover, using continuity, we see that

$\lim_{n \to \infty} [\ln(en) - \ln(n + 2)] = \lim_{n \to \infty} \ln\left(\frac{en}{n + 2} \right) = \ln\left(\lim_{n \to \infty} \frac{en}{n + 2} \right) = \ln(e) = 1 \neq 0$

Thus, we conclude from the nth term test for divergence that the

series $\sum_{n=1}^{\infty} [\ln(en) - \ln(n + 2)]$ diverges.

432. False. Since we are given that $\lim_{n \to \infty} a_n = 1$, it follows that

$\lim_{n \to \infty} \frac{2a_n}{1 + 5a_n} = \frac{2 \lim_{n \to \infty} a_n}{1 + 5 \lim_{n \to \infty} a_n} = \frac{2}{1 + 5} = \frac{2}{6} = \frac{1}{3} \neq 0$. So, we conclude from the

nth term test for divergence that the series $\sum_{n=1}^{\infty} \frac{2a_n}{1 + 5a_n}$ diverges.

433. a. The series $\sum_{n=0}^{\infty} -2\left(\frac{5}{6} \right)^{n+1}$ is a convergent geometric series with sum

$\frac{-2\left(\frac{5}{6} \right)^{0+1}}{1 - \frac{5}{6}} = \frac{-\frac{10}{6}}{\frac{1}{6}} = -10$

434. This sequence is defined in such a way that the odd-indexed terms follow one pattern while the even-indexed terms follow another. Note that the subsequence composed of the odd-indexed terms is $-2, -1, -2, -1, \ldots$ is a divergent oscillating sequence. Thus, even though the subsequence composed of the even-indexed terms (namely $1, \frac{1}{2}, \frac{1}{4}, \ldots$) converges to zero, we have identified a subsequence that does not approach the same limit value. Therefore, the original sequence must be divergent.

435. True. Observe that $a_n = n \sin\left(\frac{1}{n}\right) = \frac{\sin\left(\frac{1}{n}\right)}{\frac{1}{n}}$, $n \geq 1$. Using the fact that

$\lim\limits_{\theta \to 0} \frac{\sin\theta}{\theta} = 1$ with $\theta = \frac{1}{n}$ (which does, indeed, approach zero as n goes to

infinity), we conclude that $\lim\limits_{n \to \infty} a_n = \lim\limits_{n \to \infty} \frac{\sin\left(\frac{1}{n}\right)}{\frac{1}{n}} = 1$. Hence, the sequence converges.

Power Series Problems

We now consider infinite series whose nth term depends not only on n, but also on powers of some variable x. Precisely, we consider so-called *power series* of the form $\sum_{n=0}^{\infty} a_n (x - c)^n$. The partial sums of such a series are really just polynomial functions (in x). The largest set of values of x for which the series will converge is called the *interval of convergence*. The ratio test is used to determine the values contained within this set. The following is a characterization of the structure of the interval of convergence:

> The intervals are symmetric about c (which is the value that makes the summand equal to zero). That is, the interval of convergence is of the form $(c - R, x + R)$, where the radius R can be zero, any positive real number, or ∞.

Another important, though much less obvious, aspect of the interval of convergence is that the following operations can be performed on the power series for any x-value in the interval of convergence:

- You can take limits term for term.
- You can differentiate term for term.

- You can integrate term for term.
- You can add, subtract, or multiply in the natural way (i.e., you can treat the power series as you would a *finite* sum).

The resulting series representation of f, namely $f(x) = \sum_{n=0}^{\infty} \frac{f^{(n)}(c)}{n!}(x-c)^n$, on the interval of convergence I, is called the Taylor series representation of f centered at c.

Taylor's Theorem

Suppose f is a function possessing derivatives of all orders. Then,
$f(x) = \sum_{n=0}^{\infty} \frac{f^{(n)}(c)}{n!}(x-c)^n$ on $(c-R, c+R)$ if and only if

$$\underbrace{\lim_{n \to \infty} \frac{f^{(n+1)}(\theta)}{(n+1)!}(\theta-c)^{n+1} = 0, \text{ for all } \theta \in (c-R, c+R)}_{\substack{\text{The error in the approximation can be made arbitrarily} \\ \text{small uniformly on the interval of convergence.}}}$$

Furthermore, such a representation is unique.

Common Power Series Representations

The following are the five power series representations for some common elementary functions, together with their intervals of convergence.

1. $e^u = \sum_{n=0}^{\infty} \frac{u^n}{n!}, \ -\infty < u < \infty$

2. $\sin u = \sum_{n=0}^{\infty} \frac{(-1)^n u^{2n+1}}{(2n+1)!}, \ -\infty < u < \infty$

3. $\cos u = \sum_{n=0}^{\infty} \frac{(-1)^n u^{2n}}{(2n)!}, \ -\infty < u < \infty$

4. $\frac{1}{1-u} = \sum_{n=0}^{\infty} u^n, \ -1 < u < 1$

5. $\arctan u = \sum_{n=0}^{\infty} \frac{(-1)^n u^{2n+1}}{2n+1}, \ -1 < u < 1$

Questions

436. True or false? The power series $\sum_{n=0}^{\infty} \frac{x^n}{5^{2n}}$ converges for all values of x in the interval $[-20,20]$.

437. True or false? If the power series $\sum_{n=0}^{\infty} a_n x^n$ converges for all x in $[-1,1]$, then the series $\sum_{n=5}^{\infty} a_n x^n$ must also converge for all x in $[-1,1]$.

438. Which of the following is the interval of convergence for $\sum_{n=0}^{\infty} \frac{x^n}{2n+3}$?

   a. $[-1,1]$
   b. $[-1,1)$
   c. $(-1,1]$
   d. $(-1,1)$

439. Which of the following is the interval of convergence for $\sum_{n=0}^{\infty} \frac{x^{n+1}}{(3n+1)^2}$?

   a. $[-1,1]$
   b. $[-1,1)$
   c. $(-1,1]$
   d. $(-1,1)$

440. Which of the following is the interval of convergence for $\sum_{n=0}^{\infty} \frac{(-5)^{n+1} x^n}{2^n}$?

   a. $\left[-\frac{2}{5}, \frac{2}{5}\right]$

   b. $\left(-\frac{2}{5}, \frac{2}{5}\right)$

   c. $\left[-\frac{5}{2}, \frac{5}{2}\right]$

   d. $\left(-\frac{5}{2}, \frac{5}{2}\right)$

441. Which of the following is the interval of convergence for $\sum_{n=2}^{\infty} \sqrt[3]{n}\, x^n$?

a. $(-1,1)$
b. $[-1,1)$
c. $(-1,1]$
d. $\{0\}$

442. Which of the following is the interval of convergence for $\sum_{n=1}^{\infty} \frac{(-1)^n (x-3)^n}{4^n}$?

a. $(-\infty,7]$
b. $(-\infty,7)$
c. $(-1,7)$
d. $[-1,7]$

443. Which of the following is the interval of convergence for $\sum_{n=1}^{\infty} \frac{(3x+2)^{n+1}}{4n+1}$?

a. $\left(-1,-\frac{1}{3}\right)$

b. $\left[-1,-\frac{1}{3}\right]$

c. $\left(-1,-\frac{1}{3}\right]$

d. $\left[-1,-\frac{1}{3}\right)$

444. Which of the following is the interval of convergence for $\sum_{n=0}^{\infty} \frac{(x+1)^{2n}}{n!}$?

a. $\{-1\}$
b. $(-\infty,\infty)$
c. $(0,2)$
d. $[0,2]$

445. Which of the following is the interval of convergence for $\sum_{n=0}^{\infty} \frac{n^2 (x+1)^{2n+1}}{3^{2n-1}}$?

a. $(-10,8)$
b. $[-10,8]$
c. $[-4,2]$
d. $(-4,2)$

For Questions 446 through 449, use the known power series $\frac{1}{1-u} = \sum_{n=0}^{\infty} u^n$,

$-1 < u < 1$, with the appropriate algebraic manipulation and calculus operation, to determine a power series representation for the given function $f(x)$.

446. $f(x) = \frac{1}{1+4x}$

447. $f(x) = \frac{2}{3x}$

448. $f(x) = \frac{1}{(1+x)^2}$

449. $f(x) = \frac{x^2}{1+5x}$

For Questions 450 through 454, compute the given integral with the help of a known power series.

450. $\int_{\frac{\pi}{3}}^{\frac{\pi}{2}} \frac{\cos x}{x} \, dx$

451. $\int \frac{1}{1+x^4} \, dx$

452. $\int e^{x^3} \, dx$

453. $\int \sqrt[3]{x} \sin x \, dx$

454. $\int_0^1 \arctan\left(x^3\right) dx$

For Questions 455 through 458, determine the power series representation of the given function $f(x)$ with the specified center.

455. $f(x) = \ln x$ centered at 1

456. $f(x) = e^{2x}$ centered at $\frac{1}{2}$

457. $f(x) = x^2 e^{3x}$ centered at 0

458. $f(x) = \frac{1}{4} x^3 \sin 2x$ centered at 0

459. Which of the following is equivalent to the power series $\displaystyle\sum_{n=0}^{\infty} \frac{(-2)^n x^{5n}}{n!}$?

    a. e^{-32x^5}

    b. e^{-2x^5}

    c. e^{-2x}

    d. none of the above

460. Which of the following is the sum of the series $\displaystyle\sum_{n=0}^{\infty} \frac{\left(-\frac{25\pi^2}{36}\right)^n}{(2n)!}$?

    a. $-\frac{1}{2}$

    b. $-\frac{\sqrt{3}}{2}$

    c. $\frac{1}{2}$

    d. $\frac{\sqrt{3}}{2}$

Answers

436. True. The interval of convergence is determined by applying the ratio test, as follows:

$$\lim_{n\to\infty}\left|\frac{\frac{x^{n+1}}{5^{2(n+1)}}}{\frac{x^n}{5^{2n}}}\right|=\lim_{n\to\infty}\left|\frac{x^{n+1}}{5^{2(n+1)}}\cdot\frac{5^{2n}}{x^n}\right|=\lim_{n\to\infty}\left|\frac{x^n x}{5^{2n}\cdot5^2}\cdot\frac{5^{2n}}{x^n}\right|=\lim_{n\to\infty}\left|\frac{x}{25}\right|=\left|\frac{x}{25}\right|=\frac{|x|}{25}$$

Now, we determine the values of x that make the result of this test <1. This requires that we solve the inequality $\frac{|x|}{25}<1$, as follows:

$\frac{|x|}{25}<1$

$|x|<25$

$-25<x<25$

So, the power series converges for all x in the interval $(-25,25)$. The endpoints would need to be checked separately to determine if they should be included in this interval of convergence. But, to answer the question posed, it is sufficient to note that $[-20,20]$ is contained in $(-25,25)$, so the power series converges at all values in $[-20,20]$.

437. True. This follows because if a new series is defined by simply removing finitely many terms from a convergent series, then the new series must still converge. This is precisely the case in this scenario because $\sum_{n=5}^{\infty}a_n x^n$ is obtained by subtracting the first five terms from $\sum_{n=0}^{\infty}a_n x^n$.

438. b. The interval of convergence is determined by applying the ratio test, as follows:

$$\lim_{n\to\infty}\left|\frac{\frac{x^{n+1}}{2(n+1)+3}}{\frac{x^n}{2n+3}}\right| = \lim_{n\to\infty}\left|\frac{x^{n+1}}{2n+5}\cdot\frac{2n+3}{x^n}\right| = |x|\lim_{n\to\infty}\left|\frac{2n+3}{2n+5}\right| = |x|$$

Now, we determine the values of x that make the result of this test <1. This is given by the inequality $|x|<1$, or equivalently $-1<x<1$. Hence, the power series converges at least for all x in the interval $(-1,1)$.

The endpoints must be checked separately to determine if they should be included in the interval of convergence of the power series $\sum_{n=0}^{\infty}\frac{x^n}{2n+3}$. First, substituting $x=1$ results in the series $\sum_{n=0}^{\infty}\frac{1}{2n+3}$. Using the limit comparison test with the divergent harmonic series $\sum_{n=1}^{\infty}\frac{1}{n}$, we conclude that since $\lim_{n\to\infty}\frac{\frac{1}{2n+3}}{\frac{1}{n}} = \lim_{n\to\infty}\frac{n}{2n+3} = \frac{1}{2} > 0$, it follows that $\sum_{n=0}^{\infty}\frac{1}{2n+3}$ diverges. So, $x=1$ is not included in the interval of convergence. As for $x=-1$, we must determine if the series $\sum_{n=0}^{\infty}\frac{(-1)^n}{2n+3}$ converges. Applying the alternating series test, we note that since the sequence $\left\{\frac{1}{2n+3}\right\}_{n=0}^{\infty}$ consists of positive terms that decrease toward zero, we conclude that $\sum_{n=0}^{\infty}\frac{(-1)^n}{2n+3}$ converges. So, $x=-1$ is included in the interval of convergence.

Hence, we conclude that the interval of convergence is $[-1,1)$.

439. a. The interval of convergence is determined by applying the ratio test, as follows:

$$\lim_{n \to \infty}\left|\frac{\frac{x^{n+2}}{(3(n+1)+1)^2}}{\frac{x^{n+1}}{(3n+1)^2}}\right| = \lim_{n \to \infty}\left|\frac{x^{n+2}}{(3(n+1)+1)^2}\cdot\frac{(3n+1)^2}{x^{n+1}}\right| = |x|\lim_{n \to \infty}\left|\frac{(3n+1)^2}{(3n+4)^2}\right| = |x|$$

Now, we determine the values of x that make the result of this test <1. This is given by the inequality $|x| < 1$, or equivalently $-1 < x < 1$. Hence, the power series converges at least for all x in the interval $(-1,1)$.

The endpoints must be checked separately to determine if they should be included in the interval of convergence of the power series $\sum_{n=0}^{\infty}\frac{x^{n+1}}{(3n+1)^2}$. First, substituting $x = 1$ results in the series $\sum_{n=0}^{\infty}\frac{1}{(3n+1)^2}$.

Using the limit comparison test with the convergent p-series $\sum_{n=1}^{\infty}\frac{1}{n^2}$, we

conclude that since $\lim_{n \to \infty}\frac{\frac{1}{(3n+1)^2}}{\frac{1}{n^2}} = \lim_{n \to \infty}\frac{n^2}{(3n+1)^2} = \lim_{n \to \infty}\frac{n^2}{9n^2+6n+1} = \frac{1}{9} > 0$,

it follows that $\sum_{n=0}^{\infty}\frac{1}{(3n+1)^2}$ converges. So, $x = 1$ is included in the interval

of convergence. As for $x = -1$, we must determine if the series

$\sum_{n=0}^{\infty}\frac{(-1)^{n+1}}{(3n+1)^2}$ converges. Applying the alternating series test, we note that

since the sequence $\left\{\frac{1}{(3n+1)^2}\right\}_{n=0}^{\infty}$ consists of positive terms that decrease

toward zero, we conclude that $\sum_{n=0}^{\infty}\frac{(-1)^{n+1}}{(3n+1)^2}$ converges. So, $x = -1$ is

included in the interval of convergence.

Hence, we conclude that the interval of convergence is $[-1,1]$.

440. b. The interval of convergence is determined by applying the ratio test, as follows:

$$\lim_{n\to\infty}\left|\frac{\frac{(-5)^{n+2}\,x^{n+1}}{2^{n+1}}}{\frac{(-5)^{n+1}\,x^{n}}{2^{n}}}\right| = \lim_{n\to\infty}\left|\frac{(-5)^{n+2}\,x^{n+1}}{2^{n+1}}\cdot\frac{2^{n}}{(-5)^{n+1}\,x^{n}}\right| = |x|\lim_{n\to\infty}\left|\frac{-5}{2}\right| = \frac{5}{2}|x|$$

Now, we determine the values of x that make the result of this test <1. This is given by the inequality $\frac{5}{2}|x| < 1$, or equivalently $|x| < \frac{2}{5}$, so that $-\frac{2}{5} < x < \frac{2}{5}$. Hence, the power series converges at least for all x in the interval $\left(-\frac{2}{5},\frac{2}{5}\right)$.

The endpoints must be checked separately to determine if they should be included in the interval of convergence of the power series $\sum_{n=0}^{\infty}\frac{(-5)^{n+1}x^{n}}{2^{n}}$. First, substituting $x = -\frac{2}{5}$ results in the series

$$\sum_{n=0}^{\infty}\frac{(-5)^{n+1}\left(-\frac{2}{5}\right)^{n}}{2^{n}} = \sum_{n=0}^{\infty}\frac{(-5)^{n+1}\frac{2^{n}}{(-5)^{n}}}{2^{n}} = \sum_{n=0}^{\infty}-5,$$ which is divergent (since the

sequence of partial sums defined by $S_{n} = -5n$ goes to $-\infty$ as n gets large). Thus, $x = -\frac{2}{5}$ is not included in the interval of convergence. As for $x = \frac{2}{5}$, substituting this value of x results in the series

$$\sum_{n=0}^{\infty}\frac{(-5)^{n+1}\left(\frac{2}{5}\right)^{n}}{2^{n}} = \sum_{n=0}^{\infty}\frac{(-1)^{n+1}(5)^{n+1}\frac{2^{n}}{5^{n}}}{2^{n}} = \sum_{n=0}^{\infty}(-1)^{n+1}5,$$ which is also divergent

(since the sequence of partial sums defined by $S_{n} = (-1)^{n+1}5$ oscillates between -5 and 0 and so does not converge to a limit as n gets larger). Thus, $x = \frac{2}{5}$ is also not included in the interval of convergence.

Hence, we conclude that the interval of convergence is $\left(-\frac{2}{5},\frac{2}{5}\right)$.

441. **a.** The interval of convergence is determined by applying the ratio test, as follows:

$$\lim_{n\to\infty}\left|\frac{\sqrt[3]{n+1}\,x^{n+1}}{\sqrt[3]{n}\,x^n}\right| = |x|\lim_{n\to\infty}\sqrt[3]{\tfrac{n+1}{n}} = |x|$$

Now, we determine the values of x that make the result of this test <1. This is given by the inequality $|x| < 1$, or equivalently $-1 < x < 1$. Thus, the power series converges at least for all x in the interval $(-1,1)$.

The endpoints must be checked separately to determine if they should be included in the interval of convergence of the power series $\sum_{n=2}^{\infty}\sqrt[3]{n}\,x^n$. First, substituting $x = 1$ results in the series $\sum_{n=2}^{\infty}\sqrt[3]{n}$. This series clearly diverges because the nth term does not converge to zero as n goes to infinity; in fact, the nth term becomes arbitrarily large. So, $x = 1$ is not included in the interval of convergence. As for $x = -1$, we must determine if the series $\sum_{n=2}^{\infty}(-1)^n\sqrt[3]{n}$ converges. Here again, the nth term does not go to zero as n goes to infinity; rather, the odd-indexed terms go to negative infinity and the even-indexed terms go to plus infinity. So, $x = -1$ is also not included in the interval of convergence.

Hence, we conclude that the interval of convergence is $(-1,1)$.

442. c. The interval of convergence is determined by applying the ratio test, as follows:

$$\lim_{n\to\infty}\left|\frac{\frac{(-1)^{n+1}(x-3)^{n+1}}{4^{n+1}}}{\frac{(-1)^{n}(x-3)^{n}}{4^{n}}}\right| = \lim_{n\to\infty}\left|\frac{(-1)^{n+1}(x-3)^{n+1}}{4^{n+1}}\cdot\frac{4^{n}}{(-1)^{n}(x-3)^{n}}\right| = |x-3|\lim_{n\to\infty}\left|\frac{-1}{4}\right| = \tfrac{1}{4}|x-3|$$

Now, we determine the values of x that make the result of this test <1. This is given by the inequality $\frac{1}{4}|x-3|<1$, which is solved as follows:

$$|x-3|<4$$
$$-4<x-3<4$$
$$-1<x<7$$

Thus, the power series converges at least for all x in the interval $(-1,7)$.

The endpoints must be checked separately to determine if they should be included in the interval of convergence of the power series $\sum_{n=1}^{\infty}\frac{(-1)^{n}(x-3)^{n}}{4^{n}}$. First, substituting $x=-1$ results in the series

$$\sum_{n=1}^{\infty}\frac{(-1)^{n}(-1-3)^{n}}{4^{n}}=\sum_{n=1}^{\infty}\frac{(-1)^{n}(-4)^{n}}{4^{n}}=\sum_{n=1}^{\infty}\frac{(-1)^{n}(-1)^{n}4^{n}}{4^{n}}=\sum_{n=1}^{\infty}(-1)^{2n}=\sum_{n=1}^{\infty}1$$

This series clearly diverges because the nth term does not converge to zero as n goes to infinity; in fact, the nth term simply stays at 1. So, $x=-1$ is not included in the interval of convergence. As for $x=7$, we must determine if the series $\sum_{n=1}^{\infty}\frac{(-1)^{n}(7-3)^{n}}{4^{n}}=\sum_{n=1}^{\infty}\frac{(-1)^{n}4^{n}}{4^{n}}=\sum_{n=1}^{\infty}(-1)^{n}$ converges. Here again, the nth term does not go to zero as n goes to infinity; rather, the odd-indexed terms go to -1 and the even-indexed terms go to 1. So, $x=7$ is also not included in the interval of convergence.

Hence, we conclude that the interval of convergence is $(-1,7)$.

443. d. The interval of convergence is determined by applying the ratio test, as follows:

$$\lim_{n\to\infty}\left|\frac{\frac{(3x+2)^{n+2}}{4(n+1)+1}}{\frac{(3x+2)^{n+1}}{4n+1}}\right|=\lim_{n\to\infty}\left|\frac{(3x+2)^{n+2}}{4(n+1)+1}\cdot\frac{4n+1}{(3x+2)^{n+1}}\right|=|3x+2|\lim_{n\to\infty}\left|\frac{4n+1}{4n+5}\right|=|3x+2|$$

Now, we determine the values of x that make the result of this test <1. This is given by the inequality $|3x+2|<1$, which is solved as follows:

$|3x+2|<1$

$-1<3x+2<1$

$-3<3x<-1$

$-1<x<-\frac{1}{3}$

Hence, the power series converges at least for all x in the interval $\left(-1,-\frac{1}{3}\right)$.

    The endpoints must be checked separately to determine if they should be included in the interval of convergence of the power series $\sum_{n=1}^{\infty}\frac{(3x+2)^{n+1}}{4n+1}$. First, substituting $x=-1$ results in the series

$\sum_{n=1}^{\infty}\frac{(3(-1)+2)^{n+1}}{4n+1}=\sum_{n=1}^{\infty}\frac{(-1)^{n+1}}{4n+1}$. Applying the alternating series test, note that

since the sequence $\left\{\frac{1}{4n+1}\right\}_{n=1}^{\infty}$ consists of positive terms that decrease

toward zero, we conclude that $\sum_{n=1}^{\infty}\frac{(-1)^{n+1}}{4n+1}$ converges. So, $x=-1$ is

included in the interval of convergence. As for $x=-\frac{1}{3}$, we must

determine if the series $\sum_{n=1}^{\infty}\frac{\left(3\left(-\frac{1}{3}\right)+2\right)^{n+1}}{4n+1}=\sum_{n=1}^{\infty}\frac{1}{4n+1}$ converges. Using the

limit comparison test with the divergent harmonic series $\sum_{n=1}^{\infty}\frac{1}{n}$, we

conclude that since $\lim_{n\to\infty}\frac{\frac{1}{4n+1}}{\frac{1}{n}}=\lim_{n\to\infty}\frac{n}{4n+1}=\frac{1}{4}>0$, it follows that $\sum_{n=1}^{\infty}\frac{1}{4n+1}$

diverges. So $x=-\frac{1}{3}$ is not included in the interval of convergence.

    Hence, we conclude that the interval of convergence is $\left[-1,-\frac{1}{3}\right)$.

444. b. The interval of convergence is determined by applying the ratio test, as follows:

$$\lim_{n\to\infty}\left|\frac{\frac{(x+1)^{2(n+1)}}{(n+1)!}}{\frac{(x+1)^{2n}}{n!}}\right|=\lim_{n\to\infty}\left|\frac{(x+1)^{2(n+1)}}{(n+1)!}\cdot\frac{n!}{(x+1)^{2n}}\right|=(x+1)^2\lim_{n\to\infty}\left|\frac{1}{n+1}\right|=(x+1)^2\cdot 0=0$$

Note that this limit value is less than 1, no matter what real value of x we use. Thus, the power series converges for all x in the interval $(-\infty,\infty)$.

445. d. The interval of convergence is determined by applying the ratio test, as follows:

$$\lim_{n\to\infty}\left|\frac{\frac{(n+1)^2(x+1)^{2(n+1)+1}}{3^{2(n+1)-1}}}{\frac{n^2(x+1)^{2n+1}}{3^{2n-1}}}\right|=\lim_{n\to\infty}\left|\frac{(n+1)^2(x+1)^{2(n+1)+1}}{3^{2(n+1)-1}}\cdot\frac{3^{2n-1}}{n^2(x+1)^{2n+1}}\right|$$

$$=\lim_{n\to\infty}\left|\frac{(n+1)^2(x+1)^{2n+3}}{3^{2n+1}}\cdot\frac{3^{2n-1}}{n^2(x+1)^{2n+1}}\right|=(x+1)^2\lim_{n\to\infty}\left|\frac{(n+1)^2}{9n^2}\right|$$

$$=(x+1)^2\lim_{n\to\infty}\left|\frac{n^2+2n+1}{9n^2}\right|=\tfrac{1}{9}(x+1)^2$$

Now, we determine the values of x that make the result of this test <1. This is given by the inequality $\frac{1}{9}(x+1)^2<1$, which is solved as follows:

$\frac{1}{9}(x+1)^2<1$

$(x+1)^2<9$

$(x+1)^2-9<0$

$(x+1-3)(x+1+3)<0$

$(x-2)(x+4)<0$

$-4<x<2$

Hence, the power series converges at least for all x in the interval $(-4,2)$.

The endpoints must be checked separately to determine if they should be included in the interval of convergence of the power series $\sum_{n=0}^{\infty}\frac{n^2(x+1)^{2n+1}}{3^{2n-1}}$. First, substituting $x=-4$ results in the series

$$\sum_{n=0}^{\infty}\frac{n^2(-4+1)^{2n+1}}{3^{2n-1}}=\sum_{n=0}^{\infty}\frac{n^2(-3)^{2n+1}}{3^{2n-1}}=\sum_{n=0}^{\infty}\frac{n^2(-1)^{2n+1}3^{2n+1}}{3^{2n-1}}=\sum_{n=0}^{\infty}(-1)^{2n+1}9n^2$$

This series clearly diverges because the nth term does not converge to zero as n goes to infinity; rather, the terms go to minus infinity. So, $x=-4$ is not included in the interval of convergence. As for $x=2$, we must determine if the series $\sum_{n=0}^{\infty}\frac{n^2(2+1)^{2n+1}}{3^{2n-1}}=\sum_{n=0}^{\infty}9n^2$ converges.

Here again, the nth term does not go to zero as n goes to infinity. So, $x = 2$ is also not included in the interval of convergence.

Hence, we conclude that the interval of convergence is $(-4,2)$.

446. Observe that $f(x) = \frac{1}{1+4x} = \frac{1}{1-(-4x)}$. So, using $u = -4x$ in the known

power series $\frac{1}{1-u} = \sum_{n=0}^{\infty} u^n$, $-1 < u < 1$, yields

$f(x) = \sum_{n=0}^{\infty} (-4x)^n = \sum_{n=0}^{\infty} (-4)^n x^n$. This formula holds for any x such

that $-1 < -4x < 1$, which is equivalent to $-\frac{1}{4} < x < \frac{1}{4}$.

447. Observe that $f(x) = \frac{2}{3x} = 2\left[\frac{1}{1-(1-3x)}\right]$.

So, using $u = 1 - 3x$ in the

known power series $\frac{1}{1-u} = \sum_{n=0}^{\infty} u^n$, $-1 < u < 1$, yields $f(x) = 2\sum_{n=0}^{\infty}(1-3x)^n$.

This formula holds for any x such that $-1 < 1 - 3x < 1$, which is simplified as follows:

$-1 < 1 - 3x < 1$

$-2 < -3x < 0$

$\frac{2}{3} > x > 0$

448. First, observe that using $u = -x$ in the known power series $\frac{1}{1-u} = \sum\limits_{n=0}^{\infty} u^n$,

$-1 < u < 1$, yields $\frac{1}{1+x} = \frac{1}{1-(-x)} = \sum\limits_{n=0}^{\infty} (-x)^n = \sum\limits_{n=0}^{\infty} (-1)^n x^n$. This holds for

all $-1 < x < 1$. For all x in this interval, we can differentiate both sides of this equality to obtain:

$$\frac{d}{dx}\left(\frac{1}{1+x}\right) = -\frac{1}{(1+x)^2}$$

$$\frac{d}{dx}\sum\limits_{n=0}^{\infty} (-1)^n x^n = \sum\limits_{n=0}^{\infty} (-1)^n \frac{d}{dx}\left(x^n\right) = \sum\limits_{n=1}^{\infty} (-1)^n n x^{n-1}$$

(Note that the starting value of the index of the summation is increased by 1 upon differentiation since the first term of the series using the original starting index value would be zero, and hence would not contribute meaningfully to the sum.) The quantities in the two lines are equal. Therefore, $-\frac{1}{(1+x)^2} = \sum\limits_{n=0}^{\infty} (-1)^n n x^{n-1}$, so multiplying both

sides by -1 yields the required power series for f, namely

$f(x) = \frac{1}{(1+x)^2} = -\sum\limits_{n=0}^{\infty} (-1)^n n x^{n-1} = \sum\limits_{n=1}^{\infty} (-1)^{n+1} n x^{n-1}$. (Note that in the last

step, we were able to use linearity to bring the -1 inside the sum and combine with the summand because the series is known to converge on the interval $-1 < x < 1$.)

449. Observe that $f(x) = \frac{x^2}{1+5x} = x^2 \left[\frac{1}{1+5x} \right] = x^2 \left[\frac{1}{1-(-5x)} \right]$. So, for the

quantity in the brackets, using $u = -5x$ in the known power series

$\frac{1}{1-u} = \sum\limits_{n=0}^{\infty} u^n$, $-1 < u < 1$, yields $f(x) = x^2 \left[\sum\limits_{n=0}^{\infty} (-5x)^n \right] = x^2 \left[\sum\limits_{n=0}^{\infty} (-5)^n x^n \right]$.

This formula holds for any x such that $-1 < -5x < 1$, which is equivalent to $-\frac{1}{5} < x < \frac{1}{5}$. Now, for any such x, the series converges, so we can use linearity to bring the term inside and combine with the summand to arrive at the simplified power series $f(x) = \sum\limits_{n=0}^{\infty} (-5)^n x^{n+2}$.

450. We use the power series for cos$x$ and integrate term for term, as follows:

$$\int_{\frac{\pi}{3}}^{\frac{\pi}{2}} \frac{\cos x}{x}\, dx = \int_{\frac{\pi}{3}}^{\frac{\pi}{2}} \frac{\sum_{n=0}^{\infty}(-1)^n \frac{x^{2n}}{(2n)!}}{x}\, dx$$

$$= \int_{\frac{\pi}{3}}^{\frac{\pi}{2}} \frac{1}{x}\cdot \sum_{n=0}^{\infty}(-1)^n \frac{x^{2n}}{(2n)!}\, dx$$

$$= \int_{\frac{\pi}{3}}^{\frac{\pi}{2}} \sum_{n=0}^{\infty} \frac{1}{x}\cdot (-1)^n \frac{x^{2n}}{(2n)!}\, dx$$

$$= \sum_{n=0}^{\infty} \int_{\frac{\pi}{3}}^{\frac{\pi}{2}} (-1)^n \frac{x^{2n-1}}{(2n)!}\, dx$$

$$= \sum_{n=0}^{\infty} \frac{(-1)^n}{(2n)!} \int_{\frac{\pi}{3}}^{\frac{\pi}{2}} x^{2n-1}\, dx$$

$$= \sum_{n=0}^{\infty} \frac{(-1)^n}{(2n)!}\cdot \frac{x^{2n}}{2n}\Bigg|_{x=\frac{\pi}{3}}^{x=\frac{\pi}{2}}$$

$$= \sum_{n=0}^{\infty} \frac{(-1)^n}{2n\,(2n)!}\cdot \left[\left(\frac{\pi}{2}\right)^{2n} - \left(\frac{\pi}{3}\right)^{2n}\right]$$

451. We express the integrand as a power series that can be integrated term by term by using $u = -x^4$ in the known power series $\frac{1}{1-u} = \sum_{n=0}^{\infty} u^n$, $-1 < u < 1$, to obtain $\frac{1}{1-(-x^4)} = \sum_{n=0}^{\infty}(-x^4)^n = \sum_{n=0}^{\infty}(-1)^n x^{4n}$. This formula holds for any x such that $-1 < x^4 < 1$, which holds whenever $-1 < x < 1$. Now, substitute this formula in for the integral and compute as follows:

$$\int \frac{1}{1+x^4}\, dx = \int \sum_{n=0}^{\infty}(-1)^n x^{4n}\, dx = \sum_{n=0}^{\infty}(-1)^n \int x^{4n}\, dx = \sum_{n=0}^{\infty}(-1)^n \frac{x^{4n+1}}{4n+1} + C$$

where C is the constant of integration.

452. We express the integrand as a power series that can be integrated term by term by using $u = x^3$ in the known power series $e^u = \sum_{n=0}^{\infty} \frac{u^n}{n!}$, $-\infty < u < \infty$, to obtain $e^{x^3} = \sum_{n=0}^{\infty} \frac{(x^3)^n}{n!} = \sum_{n=0}^{\infty} \frac{x^{3n}}{n!}$. This formula holds for any real number x. Now, substitute this formula in for the integrand and compute as follows:

$$\int e^{x^3}\, dx = \int \sum_{n=0}^{\infty} \frac{x^{3n}}{n!}\, dx = \sum_{n=0}^{\infty} \frac{1}{n!}\int x^{3n}\, dx = \sum_{n=0}^{\infty} \frac{x^{3n+1} + C}{(3n+1)n!}$$

where C is the constant of integration.

453. We express the integrand as a power series that can be integrated term by term by substituting the known power series $\sin x = \sum\limits_{n=0}^{\infty} \dfrac{(-1)^n x^{2n+1}}{(2n+1)!}$,

$-\infty < x < \infty$, and combining with $\sqrt[3]{x}$, also in the integrand, to obtain

$$\sqrt[3]{x}\sin x = \sqrt[3]{x}\left[\sum_{n=0}^{\infty}\frac{(-1)^n x^{2n+1}}{(2n+1)!}\right] = \sum_{n=0}^{\infty}\frac{(-1)^n x^{\frac{1}{3}}x^{2n+1}}{(2n+1)!} = \sum_{n=0}^{\infty}\frac{(-1)^n x^{2n+\frac{4}{3}}}{(2n+1)!}$$

Now, we compute the given integral as follows:

$$\int \sqrt[3]{x}\sin x\,dx = \int \sum_{n=0}^{\infty}\frac{(-1)^n x^{2n+\frac{4}{3}}}{(2n+1)!}\,dx = \sum_{n=0}^{\infty}\frac{(-1)^n}{(2n+1)!}\int x^{2n+\frac{4}{3}}\,dx$$

$$= \sum_{n=0}^{\infty}\frac{(-1)^n}{(2n+1)!}\frac{x^{2n+\frac{7}{3}}}{\left(2n+\frac{7}{3}\right)} + C$$

454. We express the integrand as a power series that can be integrated term by term by using $u = x^3$ in the known power series $\arctan u = \sum\limits_{n=0}^{\infty}\dfrac{(-1)^n u^{2n+1}}{2n+1}$

into the integrand; this formula holds for any x for which $-1 < x^3 < 1$; this holds whenever $-1 < x < 1$. We compute as follows:

$$\int_0^1 \arctan\left(x^3\right)dx = \int_0^1 \sum_{n=0}^{\infty}\frac{(-1)^n \left(x^3\right)^{2n+1}}{2n+1}\,dx = \sum_{n=0}^{\infty}\frac{(-1)^n}{2n+1}\int_0^1 x^{6n+3}\,dx$$

$$= \sum_{n=0}^{\infty}\frac{(-1)^n}{2n+1}\cdot\frac{x^{6n+4}}{6n+4}\Bigg|_0^1 = \sum_{n=0}^{\infty}\frac{(-1)^n}{2n+1}\cdot\frac{1}{6n+4} = \sum_{n=0}^{\infty}\frac{(-1)^n}{(2n+1)(6n+4)}$$

455. In order to determine the Taylor formula representation for $f(x) = \ln x$ centered at 1, we apply the following computations:

n	$f^{(n)}(x)$	$f^{(n)}(1)$	nth Taylor coefficient $\frac{f^{(n)}(1)}{n!}$	nth term of Taylor series $\frac{f^{(n)}(1)}{n!}(x-1)^n$
0	$\ln x$	$\ln 1 = 0$	$\frac{0}{0!} = 0$	$0(x-1)^0$
1	$\frac{1}{x}$	$\frac{1}{1} = 1$	$\frac{1}{1!} = \frac{1}{1}$	$\frac{1}{1}(x-1)^1$
2	$-\frac{1}{x^2}$	$-\frac{1}{1} = -1$	$\frac{-1}{2!} = -\frac{1}{2}$	$-\frac{1}{2}(x-1)^2$
3	$\frac{2\cdot1}{x^3}$	$\frac{2\cdot1}{1} = 2!$	$\frac{2!}{3!} = \frac{1}{3}$	$\frac{1}{3}(x-1)^3$
4	$-\frac{3\cdot2\cdot1}{x^4}$	$-\frac{3\cdot2\cdot1}{1} = -3!$	$\frac{-3!}{4!} = -\frac{1}{4}$	$-\frac{1}{4}(x-1)^4$
5	$\frac{4\cdot3\cdot2\cdot1}{x^5}$	$\frac{4\cdot3\cdot2\cdot1}{1} = 4!$	$\frac{4!}{5!} = \frac{1}{5}$	$\frac{1}{5}(x-1)^5$

Ignoring the term when $n = 0$ (since it is zero), the pattern that emerges is that the nth term of the Taylor series for $f(x) = \ln x$ centered at 1 is $\frac{(-1)^{n-1}}{n}(x-1)^n$, $n \geq 1$. Hence, the Taylor series we seek is $\sum_{n=1}^{\infty} \frac{(-1)^{n-1}}{n}(x-1)^n$.

456. In order to determine the Taylor formula representation for $f(x) = e^{2x}$ centered at $\left(\frac{1}{2}\right)$, we apply the following computations:

n	$f^{(n)}(x)$	$f^{(n)}\left(\frac{1}{2}\right)$	$\dfrac{f^{(n)}\left(\frac{1}{2}\right)}{n!}$	$\dfrac{f^{(n)}\left(\frac{1}{2}\right)}{n!}\left(x - \frac{1}{2}\right)^n$
0	e^{2x}	e	$\dfrac{e}{0!}$	$\dfrac{e}{0!}\left(x - \frac{1}{2}\right)^0$
1	$2e^{2x}$	$2e$	$\dfrac{2e}{1!}$	$\dfrac{2e}{1!}\left(x - \frac{1}{2}\right)^1$
2	2^2e^{2x}	2^2e	$\dfrac{2^2e}{2!}$	$\dfrac{2^2e}{2!}\left(x - \frac{1}{2}\right)^2$
3	2^3e^{2x}	2^3e	$\dfrac{2^3e}{3!}$	$\dfrac{2^3e}{3!}\left(x - \frac{1}{2}\right)^3$
4	2^4e^{2x}	2^4e	$\dfrac{2^4e}{4!}$	$\dfrac{2^4e}{4!}\left(x - \frac{1}{2}\right)^4$
5	2^5e^{2x}	2^5e	$\dfrac{2^5e}{5!}$	$\dfrac{2^5e}{5!}\left(x - \frac{1}{2}\right)^5$

The pattern that emerges is that the nth term of the Taylor series for $f(x) = e^{2x}$ centered at $\frac{1}{2}$ is $\frac{2^n e}{n!}\left(x - \frac{1}{2}\right)^n$, $n \geq 0$. Hence, the Taylor series representation for f centered at $a = \frac{1}{2}$ is given by $\sum\limits_{n=0}^{\infty} \frac{2^n e}{n!}\left(x - \frac{1}{2}\right)^n$.

457. For this one, applying the actual Taylor formula will become tedious very quickly due to the complicated nature of the function. It is more prudent to make use of the known power series representation formula $e^u = \sum\limits_{n=0}^{\infty} \frac{u^n}{n!}$, $-\infty < u < \infty$ with $u = 3x$, and multiply term by term by x^2, as follows:

$$f(x) = x^2 e^{3x} = x^2 \sum_{n=0}^{\infty} \frac{(3x)^n}{n!} = x^2 \sum_{n=0}^{\infty} \frac{3^n x^n}{n!} = \sum_{n=0}^{\infty} \frac{3^n x^{n+2}}{n!}$$

Note that this is indeed a Taylor series representation if we identify the coefficients of x^0, x, x^2 as being zero.

458. For this one, applying the actual Taylor formula will become tedious very quickly due to the complicated nature of the function. It is more prudent to make use of the known power series representation formula

$$\sin u = \sum_{n=0}^{\infty} \frac{(-1)^n u^{2n+1}}{(2n+1)!}, \; -\infty < u < \infty \text{ with } u = 2x, \text{ and multiply term by}$$

term by $\frac{1}{4}x^3$, as follows:

$$f(x) = \frac{1}{4}x^3 \sin 2x = \frac{1}{4}x^3 \sum_{n=0}^{\infty} \frac{(-1)^n (2x)^{2n+1}}{(2n+1)!} = \frac{1}{4}x^3 \sum_{n=0}^{\infty} \frac{(-1)^n 2^{2n+1} x^{2n+1}}{(2n+1)!}$$

$$= \sum_{n=0}^{\infty} \frac{(-1)^n 2^{2n+1} x^{2n+4}}{4(2n+1)!}$$

Note that this is indeed a Taylor series representation if we identify the coefficients of appropriate powers of x as being zero.

459. b. Observe that $\sum_{n=0}^{\infty} \frac{(-2)^n x^{5n}}{n!} = \sum_{n=0}^{\infty} \frac{\left(-2x^5\right)^n}{n!}$. Using $u = 2x^5$ in the known

power series $e^u = \sum_{n=0}^{\infty} \frac{u^n}{n!}, \; -\infty < u < \infty$, shows that $\sum_{n=0}^{\infty} \frac{(-2)^n x^{5n}}{n!} = e^{-2x^5}$.

460. b. Observe that $\sum_{n=0}^{\infty} \frac{\left(-\frac{25\pi^2}{36}\right)^n}{(2n)!} = \sum_{n=0}^{\infty} \frac{(-1)^n \left(\frac{5\pi}{6}\right)^{2n}}{(2n)!}$. Using $u = \frac{5\pi}{6}$ in the known

power series $\cos u = \sum_{n=0}^{\infty} \frac{(-1)^n u^{2n}}{(2n)!}, \; -\infty < u < \infty$, shows that the given series

is equal to , which is equal to $-\frac{\sqrt{3}}{2}$.

19

Parametric and Polar Equations Problems

The rules and theorems developed for single-variable calculus for functions of the form $y = f(x)$ can be used to establish similar results for functions either defined parametrically (where the input and output are decoupled from each other and are viewed as separate functions of a third variable, or parameter) or defined using polar coordinates (r, θ) instead of Cartesian coordinates (x, y). Parametrically defined functions and functions defined using polar coordinates can be expressed equivalently using Cartesian coordinates, although the resulting expression is often more complicated in nature, so less convenient to work with. One must find a way of eliminating the parameter t in a parametric function defined by $x = x(t)$, $y = y(t)$, $a \leq t \leq b$ in order to rewrite the function using Cartesian coordinates. This is often done by solving one of the expressions for t and substituting it into the other, or by using a known identity to eliminate t in both expressions simultaneously. Expressing a polar function using Cartesian coordinates, and vice versa, requires the use of the following transformation equations.

Converting between Polar Coordinates and Cartesian Coordinates

Suppose that the Cartesian point (x,y) and the polar point (r,θ) describe the same position in the Cartesian plane. Here, $0 \leq r < \infty$, $0 \leq \theta \leq 2\pi$, and x and y are real numbers. The following equations relate x and y to r and θ:

- For converting from Cartesian to polar, use $x^2 + y^2 = r^2$, $\tan\theta = \frac{y}{x}$.
- For converting from polar to Cartesian, use $x = r\cos\theta$, $y = r\sin\theta$.

Next, we present some calculus results specific to parametric and polar functions.

Polar Calculus

Suppose that $r = f(\theta)$ is a polar function.

- If f is differentiable, then the slope of the tangent line to the curve $r = f(\theta)$ is given by

$$\frac{dy}{dx} = \frac{\frac{dr}{d\theta}\sin\theta + r\cos\theta}{\frac{dr}{d\theta}\cos\theta - r\sin\theta}$$

- The area enclosed by the graph of $r = f(\theta)$ for $a \leq \theta \leq b$ is given by $\frac{1}{2}\int_a^b \left[f(\theta)\right]^2 d\theta$.

Calculus of Parametrically Defined Functions

Consider a parametrically defined function given by $x = x(t)$, $y = y(t)$, $a \leq t \leq b$.

- If x and y are differentiable, then the slope of the tangent line to the curve is given by $\frac{dy}{dx} = \frac{\frac{dy}{dt}}{\frac{dx}{dt}}$.

- If x and y are differentiable, then the length of the curve is given by $\int_a^b \sqrt{\left(\frac{dx}{dt}\right)^2 + \left(\frac{dy}{dt}\right)^2}\, dt$.

- The area enclosed by the curve on the given interval is given by $\int_a^b y(t) \cdot \frac{dx}{dt}\, dt$.

Questions

461. The polar point $\left(4, \frac{\pi}{3}\right)$ is equivalent to which of the following points expressed using Cartesian coordinates?

   a. $\left(-2, 2\sqrt{3}\right)$

   b. $\left(2\sqrt{2}, 2\sqrt{2}\right)$

   c. $\left(2\sqrt{3}, 2\right)$

   d. $\left(2, 2\sqrt{3}\right)$

462. The Cartesian point $\left(-\sqrt{3}, 1\right)$ is equivalent to which of the following points expressed using polar coordinates?

   a. $\left(-2, \frac{5\pi}{6}\right)$

   b. $\left(-2, \frac{2\pi}{3}\right)$

   c. $\left(2, \frac{5\pi}{6}\right)$

   d. $\left(2, \frac{2\pi}{3}\right)$

463. Which of the following is an accurate description of the graph of the polar function defined by $r\sin\theta + r\cos\theta = 1$?

   a. It is a line with slope 1 passing through the point $(0,1)$.

   b. It is a line with slope -1 passing through the point $(0,1)$.

   c. It is a circle with radius 1 centered at the origin.

   d. It is a circle with radius $\frac{1}{2}$ centered at the origin.

464. Which of the following Cartesian equations is equivalent to the polar equation $r^2\cos 2\theta = 1$?

   a. $y^2 - x^2 = 1$

   b. $x^2 - y^2 = 1$

   c. $2xy = 1$

   d. $2x\sqrt{x^2 + y^2} = 1$

465. Sketch the graph of the polar function $r = \cos 3\theta,\ 0 \le \theta \le 2\pi$.

466. Sketch the graph of the polar function $r\theta = 2,\ 0 \le \theta \le 2\pi$.

467. Determine the slope of the tangent line to the polar curve $r = \cos 3\theta$, $0 \le \theta \le 2\pi$, when $\theta = \frac{\pi}{6}$.

468. Compute the area enclosed by the three-leaved rose $r = 2\cos 3\theta$.

469. Compute the area enclosed by the spiral $r = \theta$ beginning at $\theta = 0$ and ending at $\theta = \pi$.

470. Determine the slope of the tangent line to the polar curve $r = 1 - 2\cos\theta$, $0 \le \theta \le 2\pi$, when $\theta = \frac{\pi}{4}$.

471. The graph of the parametrically defined curve $x = 2 + \sin^2 t$, $y = 3 + \cos^2 t$, $0 \le t \le \frac{\pi}{2}$, is a portion of a(n) _____.
a. circle
b. line
c. parabola
d. ellipse

472. The graph of the parametrically defined curve $x = 2 + \sin t$, $y = 3 + \cos t$, $0 \le t \le 2\pi$, is a _____.
a. circle
b. line segment
c. portion of a hyperbola
d. portion of a parabola

473. Which of the following is a parametric representation for the Cartesian function $y = x^2$, $-2 \le x \le 2$?
a. $x = t, y = t^2, -2 \le t \le 2$
b. $x = \sqrt{t}, y = t^2, -2 \le t \le 2$
c. $x = 2t, y = 4t^2, 0 \le t \le 2$
d. all of the above

474. Determine the slope of the tangent line to the parametrically defined curve $x = t^2$, $y = e^{t^2}$, $t > 0$, at $t = 1$.

475. Determine the equation of the tangent line to the parametrically defined curve $x = \ln t$, $y = e^{\sqrt{t}}$, $t > 0$, at $t = 1$.

476. Which of the following integrals can be used to compute the length of the parametrically defined curve $x = e^{3t}$, $y = 2e^{4t}$, $0 \le t \le 1$?

    a. $\int_0^1 \sqrt{\left(e^{3t}\right)^2 + \left(2e^{4t}\right)^2}\, dt$

    b. $\int_0^1 \sqrt{\left(3e^{3t}\right)^2 + \left(8e^{4t}\right)^2}\, dt$

    c. $\int_0^1 \sqrt{1 + \left(2e^{4t}\right)^2}\, dt$

    d. $\int_0^1 \sqrt{1 + \left(8e^{4t}\right)^2}\, dt$

477. Determine the length of the parametrically defined curve $x = 3\sin^2 t$, $y = 3\cos^2 t$, $0 \le t \le \frac{\pi}{2}$.

478. Compute the area enclosed by the parametrically defined curve $x = \sqrt{3}\sin 2t$, $y = \sqrt{5}\cos 2t$, $0 \le t \le \pi$.

479. Sketch the graph of the parametrically defined curve $x = t^2$, $y = t^4 + 1$, $-3 \le t \le 0$.

480. Sketch the graph of the parametrically defined curve $x = 2\cos t$, $y = 3\sin^2 t$, $\frac{\pi}{2} \le t \le \pi$.

Answers

461. **d.** Here, $r = 4$, $\theta = \frac{\pi}{3}$. So, using the transformation equations

$x = r\cos\theta$, $y = r\sin\theta$ yields $x = 4\cos\left(\frac{\pi}{3}\right) = 4\left(\frac{1}{2}\right) = 2$,

$y = 4\sin\left(\frac{\pi}{3}\right) = 4\left(\frac{\sqrt{3}}{2}\right) = 2\sqrt{3}$. So, the equivalent point expressed in

Cartesian coordinates is $\left(2, 2\sqrt{3}\right)$.

462. **c.** Here, $x = -\sqrt{3}$, $y = 1$. So, noting that this point is in quadrant II and using the transformation equations $x^2 + y^2 = r^2$, $\tan\theta = \frac{y}{x}$ yields the following:

$\left(-\sqrt{3}\right)^2 + 1^2 = r^2$, so $r = 2$

$\tan\theta = \frac{1}{-\sqrt{3}} = \frac{\frac{1}{2}}{\frac{-\sqrt{3}}{2}}$, so $\theta = \frac{5\pi}{6}$

So, the equivalent point expressed in polar coordinates is $\left(2, \frac{5\pi}{6}\right)$.

463. **b.** Apply the transformation equations $x = r\cos\theta$, $y = r\sin\theta$ in the given equation $r\sin\theta + r\cos\theta = 1$ to obtain the equivalent Cartesian equation $y + x = 1$, or equivalently $y = -x + 1$. This is a line with slope -1 passing through the point $(0,1)$.

464. **b.** Applying the double-angle formula $\cos 2\theta = \cos^2\theta - \sin^2\theta$ in the original equation, simplifying, and applying the transformation equations $x = r\cos\theta$, $y = r\sin\theta$, yields:

$r^2\cos 2\theta = 1$

$r^2\cos^2\theta - r^2\sin^2\theta = 1$

$x^2 - y^2 = 1$

465.

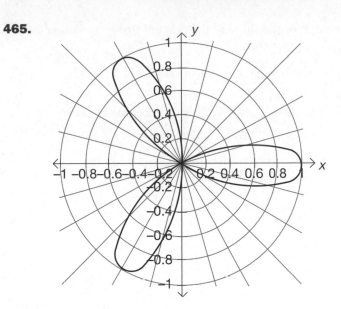

466.

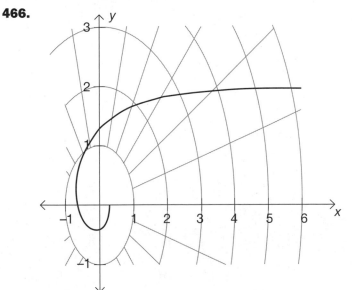

467. The general formula for the slope of the tangent line to the polar curve $r = \cos 3\theta$, $0 \leq \theta \leq 2\pi$, is given by

$$\frac{dy}{dx} = \frac{\frac{dr}{d\theta}\sin\theta + r\cos\theta}{\frac{dr}{d\theta}\cos\theta - r\sin\theta} = \frac{(-3\sin 3\theta)\sin\theta + (\cos 3\theta)\cos\theta}{(-3\sin 3\theta)\cos\theta - (\cos 3\theta)\sin\theta}$$

Evaluating this expression when $\theta = \frac{\pi}{6}$ yields the slope of the specific tangent line of interest, as follows:

$$\frac{dy}{dx} = \frac{-3\sin\left(3\cdot\frac{\pi}{6}\right)\sin\left(\frac{\pi}{6}\right) + \cos\left(3\cdot\frac{\pi}{6}\right)\cos\left(\frac{\pi}{6}\right)}{-3\sin\left(3\cdot\frac{\pi}{6}\right)\cos\left(\frac{\pi}{6}\right) - \cos\left(3\cdot\frac{\pi}{6}\right)\sin\left(\frac{\pi}{6}\right)} = \frac{-3(1)\left(\frac{1}{2}\right) + 0}{-3(1)\left(\frac{\sqrt{3}}{2}\right) - 0} = \frac{1}{\sqrt{3}} = \frac{\sqrt{3}}{3}$$

468. One complete tracing of the polar curve $r = 2\cos 3\theta$ occurs by starting at $\theta = 0$ and ending at $\theta = \frac{2\pi}{3}$, as follows:

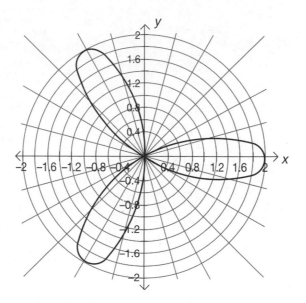

The integral used to compute the area enclosed by this three-leaved rose is given by

$$\frac{1}{2}\int_0^{\frac{2\pi}{3}}\left[2\cos 3\theta\right]^2 d\theta = 2\int_0^{\frac{2\pi}{3}}\cos^2 3\theta\, d\theta = 2\int_0^{\frac{2\pi}{3}}\frac{1+\cos 6\theta}{2}d\theta$$

$$= 1\left[\int_0^{\frac{2\pi}{3}}1\,d\theta + \int_0^{\frac{2\pi}{3}}\cos 6\theta\, d\theta\right] = \int_0^{\frac{2\pi}{3}}1\,d\theta + \int_0^{\frac{2\pi}{3}}\cos 6\theta\, d\theta$$

(Note that the double-angle formula $\cos^2 u = \frac{1+\cos 2u}{2}$ is used.) The first integral is easily computed and the second is computed using a u-substitution with $u = 6\theta$. Doing so yields:

Area $= \int_0^{\frac{2\pi}{3}}1\,d\theta + \int_0^{\frac{2\pi}{3}}\cos 6\theta\, d\theta = \theta\Big|_0^{\frac{2\pi}{3}} + \frac{1}{6}\sin 6\theta\Big|_0^{\frac{2\pi}{3}} = \left(\frac{2\pi}{3}-0\right)+\frac{1}{6}(0-0)$

$= \frac{2\pi}{3}$ units^2

469. A sketch of the polar curve $r = \theta$ starting at $\theta = 0$ and ending at $\theta = \pi$ is as follows:

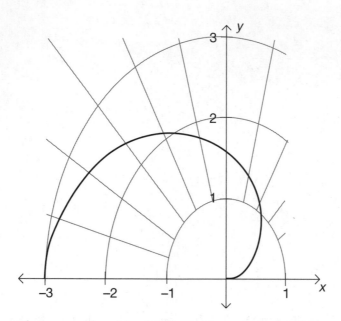

The area enclosed by this region is computed as follows:

$$\frac{1}{2} \int_0^\pi \theta^2 \, d\theta = \frac{1}{2} \cdot \frac{\theta^3}{3} \bigg|_0^\pi = \frac{\pi^3}{6} \text{ units}^2$$

470. In order to write the equation of the tangent line, we need the slope and the point on the curve that occur when $\theta = \frac{\pi}{4}$. The general formula for the slope of the tangent line to the polar curve $r = 1 - 2\cos\theta$, $0 \le \theta \le 2\pi$, is given by

$$\frac{dy}{dx} = \frac{\frac{dr}{d\theta}\sin\theta + r\cos\theta}{\frac{dr}{d\theta}\cos\theta - r\sin\theta} = \frac{(2\sin\theta)\sin\theta + (1 - 2\cos\theta)\cos\theta}{(2\sin\theta)\cos\theta - (1 - 2\cos\theta)\sin\theta}$$

Evaluating this expression when $\theta = \frac{\pi}{4}$ yields the slope of the specific tangent line of interest, as follows:

$$\frac{dy}{dx} = \frac{2\sin\frac{\pi}{4}\sin\frac{\pi}{4} + \left(1 - 2\cos\frac{\pi}{4}\right)\cos\frac{\pi}{4}}{2\sin\frac{\pi}{4}\cos\frac{\pi}{4} - \left(1 - 2\cos\frac{\pi}{4}\right)\sin\frac{\pi}{4}} = \frac{2\left(\frac{\sqrt{2}}{2}\right)\left(\frac{\sqrt{2}}{2}\right) + \left(1 - 2\cdot\frac{\sqrt{2}}{2}\right)\frac{\sqrt{2}}{2}}{2\left(\frac{\sqrt{2}}{2}\right)\left(\frac{\sqrt{2}}{2}\right) - \left(1 - 2\cdot\frac{\sqrt{2}}{2}\right)\frac{\sqrt{2}}{2}}$$

$$= \frac{1 + (1 - \sqrt{2})\left(\frac{\sqrt{2}}{2}\right)}{1 - (1 - \sqrt{2})\left(\frac{\sqrt{2}}{2}\right)} = \frac{\frac{2 + \sqrt{2} - 2}{2}}{\frac{2 + \sqrt{2} + 2}{2}} = \frac{\sqrt{2}}{4 - \sqrt{2}} = \frac{\sqrt{2}(4 + \sqrt{2})}{(4 - \sqrt{2})(4 + \sqrt{2})} = \frac{4\sqrt{2} + 2}{14} = \frac{2\sqrt{2} + 1}{7}$$

Next, note that the polar coordinates of the point of tangency (i.e., the point on the curve when $\theta = \frac{\pi}{4}$) are $r = 1 - 2\cos\frac{\pi}{4} = 1 - 2\left(\frac{\sqrt{2}}{2}\right) = 1 - \sqrt{2}$. The corresponding Cartesian coordinates are obtained using the transformation equations $x = r\cos\theta$, $y = r\sin\theta$, as follows:

$$x = r\cos\theta = \left(1 - \sqrt{2}\right)\cos\frac{\pi}{4} = \left(1 - \sqrt{2}\right)\frac{\sqrt{2}}{2} = \frac{\sqrt{2}}{2} - 1$$

$$y = r\sin\theta = \left(1 - \sqrt{2}\right)\sin\frac{\pi}{4} = \left(1 - \sqrt{2}\right)\frac{\sqrt{2}}{2} = \frac{\sqrt{2}}{2} - 1$$

Summarizing, the slope of the tangent line is $m = \frac{2\sqrt{2} + 1}{7}$ and a point through which the line passes is $\left(\frac{\sqrt{2}}{2} - 1, \frac{\sqrt{2}}{2} - 1\right)$. Hence, using the point-slope formula for the equation of a line, we conclude that the equation of the desired tangent line is given by

$$y - \left(\frac{\sqrt{2}}{2} - 1\right) = \frac{2\sqrt{2} + 1}{7}\left(x - \left(\frac{\sqrt{2}}{2} - 1\right)\right).$$

471. b. The given parametric equations are equivalent to $x - 2 = \sin^2 t$, $y - 3 = \cos^2 t$, $0 \leq t \leq \frac{\pi}{2}$. We can eliminate the parameter t by using the trigonometric identity $\sin^2 \theta + \cos^2 \theta = 1$. Doing so yields the equivalent Cartesian equation $(x - 2) + (y - 3) = 1$, or equivalently $y = -x + 6$; this is a line. The given curve is the portion of this line starting at the point $(2, 4)$ (when $t = 0$) and ending at the point $(3, 3)$ (when $t = \frac{\pi}{2}$).

472. a. The given parametric equations are equivalent to $x - 2 = \sin t$, $y - 3 = \cos t$, $0 \leq t \leq 2\pi$. We can eliminate the parameter t by using the trigonometric identity $\sin^2 \theta + \cos^2 \theta = 1$. Doing so yields, upon squaring both sides of both equations, the equivalent Cartesian equation $(x - 2)^2 + (y - 3)^2 = 1$. This is the equation of a circle with radius 1 centered at $(2, 3)$.

473. a. The natural way to parameterize a function of the form $y = f(x)$, $a \leq x \leq b$, is to identify the input variable x as the parameter t. Doing so yields the equivalent parametric form $x = t$, $y = f(t)$, $a \leq t \leq b$. Applying this approach to parameterize $y = x^2$, $-2 \leq x \leq 2$, results in $x = t$, $y = t^2$, $-2 \leq t \leq 2$.

474. The general formula for the slope of the tangent line to the parametrically defined curve $x = t^2$, $y = e^{t^2}$, $t > 0$ is given by $\frac{dy}{dx} = \frac{\frac{dy}{dt}}{\frac{dx}{dt}} = \frac{2te^{t^2}}{2t} = e^{t^2}$, so that the slope of the tangent line at $t = 1$ is $e^{t^2} = e$.

475. In order to write the equation of the tangent line, we need the slope and the point on the curve that occur when $t = 1$. The general formula for the slope of the tangent line to the parametrically defined curve $x = \ln t$, $y = e^{\sqrt{t}}$, $t > 0$ is given by $\frac{dy}{dx} = \frac{\frac{dy}{dt}}{\frac{dx}{dt}} = \frac{\frac{e^{\sqrt{t}}}{2\sqrt{t}}}{\frac{1}{t}} = \frac{\sqrt{t}e^{\sqrt{t}}}{2}$. So, the slope of the tangent line at $t = 1$ is $\frac{\sqrt{1}e^{\sqrt{1}}}{2} = \frac{e}{2}$. Also, the coordinates of the point on the curve when $t = 1$ are $x = \ln 1 = 0$, $y = e^{\sqrt{1}} = e$. Summarizing, the slope of the desired tangent line is $\frac{e}{2}$ and a point through which this line passes is $(0, e)$. Hence, the point-slope formula for the equation of this tangent line is $y - e = \frac{e}{2}(x - 0)$, or equivalently $y = \frac{e}{2}x + e$.

476. b. Note that $\frac{dx}{dt} = 3e^{3t}$, $\frac{dy}{dt} = 8e^{4t}$. So, applying the formula for the length of a parametrically defined curve, we conclude that the length of the parametrically defined curve $x = e^{3t}$, $y = 2e^{4t}$, $0 \leq t \leq 1$, is given by
$$\int_0^1 \sqrt{\left(3e^{3t}\right)^2 + \left(8e^{4t}\right)^2}\, dt.$$

477. We apply the formula $\int_a^b \sqrt{\left(\frac{dx}{dt}\right)^2 + \left(\frac{dy}{dt}\right)^2}\, dt$ with $x = 3\sin^2 t$, $y = 3\cos^2 t$, $a = 0$, and $b = \frac{\pi}{2}$ to determine the length of the given curve. Using the double-angle formula $\sin 2\theta = 2\sin\theta\cos\theta$, we have

$\frac{dx}{dt} = 3\cdot(2\sin t \cos t) = 3\sin 2t$

$\frac{dy}{dt} = -3\cdot(2\cos t \sin t) = -3\sin 2t$

So, the length of the given curve is

$$\int_0^{\frac{\pi}{2}} \sqrt{(3\sin 2t)^2 + (-3\sin 2t)^2}\, dt = \int_0^{\frac{\pi}{2}} \sqrt{18\sin^2 2t}\, dt = \sqrt{18} \int_0^{\frac{\pi}{2}} \sin 2t\, dt$$

$$= \frac{3\sqrt{2}}{-2} \cos 2t \Big|_0^{\frac{\pi}{2}} = -\frac{3\sqrt{2}}{2}[\cos\pi - \cos 0]$$

$$= -\frac{3\sqrt{2}}{2}[-2] = 3\sqrt{2} \text{ units}$$

(Note that technically $\sqrt{\sin^2 2t} = |\sin 2t|$, but since $\sin 2t$ is nonnegative on the interval of integration, we can discard the absolute value. Also, in computing the antiderivative of $\sin 2t$, we inherently used the u-substitution $u = 2t$.)

478. The area enclosed by the parametrically defined curve $x = \sqrt{3}\sin 2t$, $y = \sqrt{5}\cos 2t$, $0 \le t \le \pi$, is given by

$$\int_0^\pi \left(\sqrt{5}\cos 2t\right)\left(\sqrt{3}\sin 2t\right)'\, dt = \int_0^\pi \left(\sqrt{5}\cos 2t\right)\left(2\sqrt{3}\cos 2t\right) dt$$

$$= 2\sqrt{15} \int_0^\pi \cos^2(2t)\, dt = 2\sqrt{15} \int_0^\pi \frac{1 + \cos(4t)}{2}\, dt$$

$$= \sqrt{15} \int_0^\pi [1 + \cos(4t)]\, dt$$

$$= \sqrt{15}\left[t + \tfrac{1}{4}\sin(4t)\right]\Big|_0^\pi$$

$$= \sqrt{15}[(\pi + 0) - (0 + 0)] = \pi\sqrt{15} \text{ units}^2$$

479. First, determine the coordinates of a few points to help determine the direction in which the curve is traced:

t	x	y
−3	9	82
−2	4	17
−1	1	2
0	0	1

The graph is as follows:

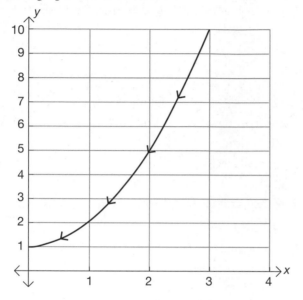

480. First, determine the coordinates of a few points to help determine the direction in which the curve is traced:

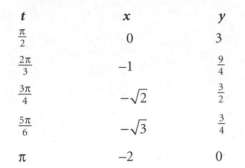

t	x	y
$\frac{\pi}{2}$	0	3
$\frac{2\pi}{3}$	-1	$\frac{9}{4}$
$\frac{3\pi}{4}$	$-\sqrt{2}$	$\frac{3}{2}$
$\frac{5\pi}{6}$	$-\sqrt{3}$	$\frac{3}{4}$
π	-2	0

The graph is as follows:

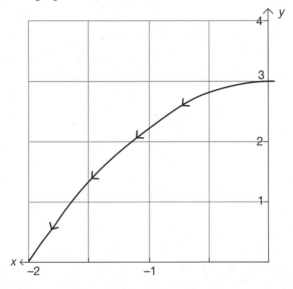

20

Common Calculus Errors

For each of the following scenarios, clearly identify the nature of the error and how to fix it.

Questions

481. $\lim\limits_{x\to\infty}\dfrac{5x^2+3x+1}{4x^2+3}=\dfrac{5(\infty)^2+3(\infty)+1}{4(\infty)^2+3}=\dfrac{\infty}{\infty}=1.$

482. The graph of the piecewise-defined function $f(x)=\begin{cases}1-(x-3)^2, & x\neq 3,\\ \quad 2, & x=3\end{cases}$ is given here:

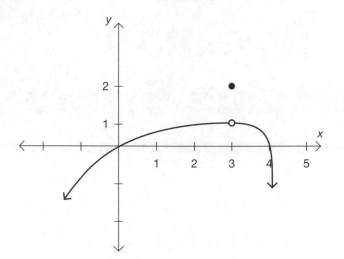

Then, $\lim\limits_{x\to 3}f(x)=f(3)=2.$

483. Let $f(x)=x^2$. By definition, we have
$$\lim_{h\to 0}\frac{f(x+h)-f(x)}{h}=\lim_{h\to 0}\frac{(x^2+h)-x^2}{h}=\lim_{h\to 0}\frac{h}{h}=\lim_{h\to 0}1=1$$
Hence, we conclude that $f'(x)=1.$

484. To find the equation of the tangent line to the graph of $f(x)=e^{3x}$ when $x=\ln(2)$, we must identify the slope of the line and the point of tangency at this x-value. Since $f(\ln 2)=e^{3\ln 2}=e^{\ln 2^3}=8$, it follows that the point of tangency is $(\ln(2),8)$. Also, the slope is computed using the derivative, namely $f'(x)=3e^{3x}$. Thus, using the point-slope form for the equation of a line, we conclude that the equation of the desired tangent line is $y-8=3e^{3x}(x-\ln 2).$

485. $\dfrac{d}{dx}x^3\cos 5x=\left(\dfrac{d}{dx}x^3\right)\left(\dfrac{d}{dx}\cos 5x\right)=\left(3x^2\right)(-5\sin 5x)=-15x^2\sin 5x$

486. $\dfrac{d}{dx}\left(\dfrac{5x^4-3x+3}{6x^3+1}\right) = \dfrac{\frac{d}{dx}\left(5x^4-3x+3\right)}{\frac{d}{dx}\left(6x^3+1\right)} = \dfrac{20x^3-3}{18x^2}$

487. $\dfrac{d}{dx}\ln\left(2-3e^x\right) = \dfrac{1}{2-3e^x}$

488. The derivative of the implicitly defined function whose equation is
$y = \sin(x^2 y^3)$ is given by $\dfrac{dy}{dx} = \cos\left(x^2 y^3\right)\cdot\left(x^2\cdot 3y^2 + y^3\cdot 2x\right)$

$= \left(3x^2 y^2 + 2xy^3\right)\cos\left(x^2 y^3\right)\cdot$

489. $\displaystyle\int x\cos x\,dx = \int x\,dx\cdot\int\cos x\,dx = \dfrac{x^2}{2}\cdot\sin x + C$

490. In order to compute $\displaystyle\int\ln(2x+1)\,dx$, first make the following
substitution:
$w = 2x+1$
$dw = 2dx \implies \tfrac{1}{2}dw = dx$
Then, we have $\displaystyle\int\ln(2x+1)\,dx = \tfrac{1}{2}\int\ln w\,dw = \dfrac{1}{2w} + C = \dfrac{1}{2(2x+1)} + C$.

491. $\dfrac{d}{dx}\displaystyle\int_2^{x^2}\sqrt[3]{5-t^2}\,dt = \sqrt[3]{5-\left(x^2\right)^2} = \sqrt[3]{5-x^4}$

492. $\dfrac{d}{dx}\displaystyle\int_{\sqrt{\pi}}^{e^x}\cos\left(t^4+1\right)dt = \cos\left(\left(e^x\right)^4+1\right)\cdot e^x - \cos\left(\left(\sqrt{\pi}\right)^4+1\right)$

$= e^x\cos\left(e^{4x}+1\right) - \cos\left(\pi^2+1\right)$

493. Observe that $\displaystyle\int_0^{\frac{3\pi}{2}}\sin x\,dx = -\cos x\Big|_0^{\frac{3\pi}{2}} = \left(-\cos\tfrac{3\pi}{2}\right)-\left(-\cos 0\right) = 0+1 = 1$.
So, since the resulting value of the integral is positive, the integral must
represent the area of the region bounded between the curve $y = \sin x$
and the x-axis between $x = 0$ and $x = \frac{3\pi}{2}$.

494. Consider the following region:

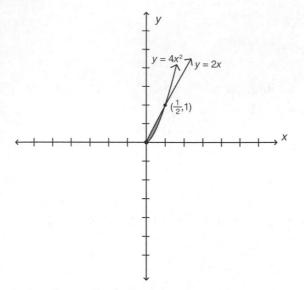

Using the method of washers, the volume of the solid obtained by revolving this region around the x-axis is given by the integral

$$\pi \int_{0}^{\frac{1}{2}} \left[4x^2 - 2x \right]^2 dx.$$

495. Consider the following region:

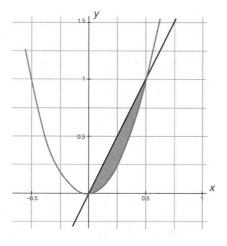

Using horizontal slices, the area of this region is given by the integral

$$\int_{0}^{8} \left[x^3 - 4\sqrt{2x} \right] dx.$$

496. Suppose that $\{a_n\}$ is a convergent sequence of nonnegative real numbers. Then, the series $\sum\limits_{n=1}^{\infty} a_n$ must also converge.

497. To determine the interval of convergence for the power series $\sum\limits_{n=0}^{\infty} \frac{(2x)^n}{3n+1}$, applying the ratio test yields:

$$\lim_{n \to \infty} \frac{\frac{(2x)^{n+1}}{3(n+1)+1}}{\frac{(2x)^n}{3n+1}} = \lim_{n \to \infty} \frac{(2x)^{n+1}}{3(n+1)+1} \cdot \frac{3n+1}{(2x)^n} = 2x \lim_{n \to \infty} \frac{3n+1}{3n+4} = 2x$$

Thus, imposing the restriction that the outcome be less than 1, we infer that the x-values for which the series converges satisfy the inequality $2x < 1$, or equivalently $x < \frac{1}{2}$. So, the interval of convergence is $\left(-\infty, \frac{1}{2}\right)$.

498. The series $\sum\limits_{n=0}^{\infty} \frac{1}{3n+1}$ converges because the nth term goes to zero as n goes to infinity.

499. Suppose that the terms of the sequence $\{c_n\}$ satisfy the inequality $-1 - \frac{1}{n} \le c_n \le 1 + \frac{1}{n}$, for every n. Since the outer sequences $\left\{-1 - \frac{1}{n}\right\}$ and $\left\{1 + \frac{1}{n}\right\}$ both converge, we conclude from the squeeze theorem that $\{c_n\}$ must also converge.

500. Since the function $f(x) = |x - 1|$ is continuous at every real number x, it is also differentiable at every real number x.

501. $\int_{-2}^{1} \frac{1}{x^2}\,dx = -\frac{1}{x}\Big|_{-2}^{1} = -1 - \left(-\frac{1}{-2}\right) = -\frac{3}{2}$

Answers

481. There are two related errors within this computation. First, "∞" cannot be treated as if it were a real number. Specifically, the first step in which the symbol ∞ is substituted into the variable is nonsensical. The second error is in claiming that $\frac{\infty}{\infty} = 1$. In actuality, a limit that results in behavior like $\frac{\infty}{\infty}$ is said to be *indeterminate* in the sense that one cannot deduce the actual value of the limit when the expression is in its current form. The correct approach is to multiply both the numerator and the denominator by the reciprocal of the term of highest degree in the denominator, namely $4x^2$. Doing so yields

$$\lim_{x \to \infty} \frac{5x^2 + 3x + 1}{4x^2 + 3} = \lim_{x \to \infty} \frac{5x^2 + 3x + 1}{4x^2 + 3} \cdot \frac{\frac{1}{4x^2}}{\frac{1}{4x^2}} = \lim_{x \to \infty} \frac{\frac{5}{4} + \frac{3}{4x} + \frac{1}{4x^2}}{1 + \frac{3}{4x^2}}$$

$$= \frac{\frac{5}{4} + \lim_{x \to \infty} \frac{3}{4x} + \lim_{x \to \infty} \frac{1}{4x^2}}{1 + \lim_{x \to \infty} \frac{3}{4x^2}} = \frac{5}{4}$$

482. The error is assuming that the function is continuous at $x = 3$. It is not continuous because of the hole in the graph. Hence, the *limit* as x approaches 3 from either side is the real number to which the y-values on the graph become arbitrarily close. This value is 1, not 2.

483. The error is a typical computational one, namely that $f(x + h) = x^2 + h$. This is not the case. Rather, $f(x + h) = (x + h)^2 = x^2 + 2xh + h^2$. Using this in the computation leads to the correct result, as follows:

$$\lim_{h \to 0} \frac{f(x + h) - f(x)}{h} = \lim_{h \to 0} \frac{(x^2 + 2xh + h^2) - x^2}{h} = \lim_{h \to 0} \frac{2xh + h^2}{h}$$

$$= \lim_{h \to 0} \frac{h(2x + h)}{h} = \lim_{h \to 0} (2x + h) = 2x$$

Hence, we conclude that $f'(x) = 2x$.

484. Most of what is provided is correct, with the exception of the very last conclusion. You must evaluate the derivative *at $x = \ln(2)$* and substitute this real number in for the slope. Specifically, $f'(\ln 2) = 3e^{3\ln 2} = 3e^{\ln 2^3} = 3(8) = 24$. The correct conclusion is that the equation of the desired tangent line is $y - 8 = 24(x - \ln 2)$.

485. The derivative of a product is *not* the product of the derivatives. The correct derivative is as follows:

$$\frac{d}{dx} x^3 \cos 5x = x^3 \left(\frac{d}{dx} \cos 5x \right) + \cos 5x \left(\frac{d}{dx} x^3 \right)$$

$$= x^3 (-5 \sin 5x) + \cos 5x (3x^2) = -5x^3 \sin 5x + 3x^2 \cos 5x$$

486. The derivative of a quotient is *not* the quotient of the derivatives. The correct unsimplified derivative is as follows:

$$\frac{d}{dx}\left(\frac{5x^4-3x+3}{6x^3+1}\right)=\frac{\left(6x^3+1\right)\frac{d}{dx}\left(5x^4-3x+3\right)-\left(5x^4-3x+3\right)\frac{d}{dx}\left(6x^3+1\right)}{\left(6x^3+1\right)^2}$$

$$=\frac{\left(6x^3+1\right)\left(20x^3-3\right)-\left(5x^4-3x+3\right)\left(18x^2\right)}{\left(6x^3+1\right)^2}$$

487. The chain rule should have been used to compute this derivative. Precisely, $\frac{d}{dx}\ln(u(x))=\frac{1}{u(x)}\cdot u'(x)$. Applying this yields $\frac{d}{dx}\ln\left(2-3e^x\right)$

$$=\frac{1}{2-3e^x}\left(-3e^x\right).$$

488. Given that the function is implicitly defined *and* we seek $\frac{dy}{dx}$, it is understood that y is taken to be a function of x; there is simply no explicit formula for it. Thus, when differentiating both sides of the given equation with respect to x, the chain rule must be used whenever an expression involving y is encountered. The correct implicit differentiation is as follows:

$$\frac{dy}{dx}=\cos\left(x^2y^3\right)\cdot\left(x^2\cdot3y^2+\frac{d}{dx}+y^3\cdot2x\right)=3x^2y^2\cos\left(x^2y^3\right)\frac{dy}{dx}$$

$$+2xy^3\cos\left(x^2y^3\right)$$

$$\frac{dy}{dx}-3x^2y^2\cos\left(x^2y^3\right)\frac{dy}{dx}=2xy^3\cos\left(x^2y^3\right)$$

$$\frac{dy}{dx}\left[1-3x^2y^2\cos\left(x^2y^3\right)\right]=2xy^3\cos\left(x^2y^3\right)$$

$$\frac{dy}{dx}=\frac{2xy^3\cos\left(x^2y^3\right)}{1-3x^2y^2\cos\left(x^2y^3\right)}$$

489. The integral of a product is not the product of the integrals. Integration by parts must be used to compute $\int x\cos x\,dx$. Indeed, we apply the integration by parts formula $\int u\,dv=uv-\int v\,du$ with the following choices of u and v, along with their differentials:

$u=x \qquad dv=\cos x\,dx$

$du=dx \qquad v=\int\cos x\,dx=\sin x$

Applying the integration by parts formula yields:
$$\int x\cos x\,dx=x\sin x-\int\sin x\,dx$$

$$=x\sin x-(-\cos x)+C=x\sin x+\cos x+C$$

490. The substitution portion of the computation is correct, but the computation of $\int \ln w\,dw$ is incorrect. Integration by parts should be used to compute this integral; the error made in the given computation was that the *differentiation* formula for $\ln(w)$ was used rather than the *antiderivative*. Picking up the computation from that point, we apply the integration by parts formula $\int u\,dv = uv - \int v\,du$ with the following choices of u and v, along with their differentials:

$u = \ln w \qquad dv = dw$

$du = \frac{1}{w}dw \qquad v = \int 1\,dw = w$

Applying the integration by parts formula yields:

$\int \ln w\,dw = w\ln w - \int w\left(\frac{1}{w}\right)dw = w\ln w - \int 1\,dw = w\ln w - w + C$

Finally, substituting $w = 2x + 1$ into the formula yields the desired antiderivative, namely, $\int \ln(2x+1)\,dx = \frac{1}{2}\int \ln w\,dw = \frac{1}{2}[(2x+1)$

$\ln(2x+1) - (2x+1) + C]\,(2x+1)\ln(2x+1) - (2x+1) + C$

491. When applying the fundamental theorem of calculus when the upper limit is a function of x rather than just x, you must use the chain rule. Hence, the correct computation is:

$\frac{d}{dx}\int_2^{x^2}\sqrt[3]{5-t^2}\,dt = \sqrt[3]{5-\left(x^2\right)^2}\cdot\frac{d}{dx}\left(x^2\right) = 2x\sqrt[3]{5-\left(x^2\right)^2} = 2x\sqrt[3]{5-x^4}$

492. When applying the fundamental theorem of calculus, if the lower limit is constant, it does not enter into the actual derivative formula. Thus, the correct computation is:

$\frac{d}{dx}\int_{\sqrt{\pi}}^{e^x}\cos\left(t^4+1\right)dt = \cos\left(\left(e^x\right)^4+1\right)\cdot e^x = e^x\cos\left(e^{4x}+1\right)$

493. Not all positive integrals represent areas. In fact, if any portion of the function $y = f(x)$ is below the x-axis on (a,b), then simply computing $\int_a^b f(x)\,dx$ will not result in the area. This is an accurate characterization of the present situation. Indeed, note that the graph of $y = \sin x$ on the interval $\left(0, \frac{3\pi}{2}\right)$ is given by the following:

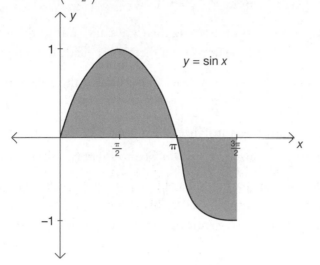

The actual area of the region bounded between $y = \sin x$ and the x-axis between $x = 0$ and $x = \frac{3\pi}{2}$ is given by:

$$\int_0^{\frac{3\pi}{2}} \sin x\,dx = \int_0^{\pi} \sin x\,dx - \int_{\pi}^{\frac{3\pi}{2}} \sin x\,dx = -\cos x\Big|_0^{\pi} - \left[-\cos x\right]\Big|_{\pi}^{\frac{3\pi}{2}}$$

$$= (-\cos \pi + \cos 0) + \left(\cos \frac{3\pi}{2} - \cos \pi\right) = (1+1) + (0+1) = 3 \text{ units}^2$$

494. This integral formula is not the one used in the washer method. Specifically, each of the two radii, namely $4x^2$ and $2x$, should be squared separately. The correct integral to use to compute this volume is

$$\pi \int_0^{\frac{1}{2}} \left[\left(4x^2\right)^2 - \left(2x\right)^2\right] dx.$$

495. The mistake is not expressing the integrand in terms of y, which must be the case when using horizontal slices. The two equations in terms of y are as follows:

$$y = x^3 \Rightarrow x = \sqrt[3]{y}$$

$$y = 4\sqrt{2x} \Rightarrow \sqrt{2x} = \frac{y}{4} \Rightarrow 2x = \left(\frac{y}{4}\right)^2 \Rightarrow x = \frac{y^2}{32}$$

Using these functions and subtracting the right curve minus the left curve to get the height of the horizontal rectangles yields the integral formula $\int_0^8 \left[\sqrt[3]{y} - \frac{y^2}{32}\right] dy$ that can be used to compute the area of the given region.

496. This is false in a big way! For instance, the nth term in any constant series $\sum_{n=0}^{\infty} c$, where c is nonzero, is certainly convergent (namely to the constant c itself), but the partial sums of the series go to positive infinity (if $c > 0$) or negative infinity (if $c < 0$). This statement is false even if the nth term goes to zero, as seen by the divergent harmonic series $\sum_{n=1}^{\infty} \frac{1}{n}$.

497. This big mistake is forgetting the absolute value around the expression of which we are taking the limit. The correct implementation of the ratio test is as follows:

$$\lim_{n\to\infty} \left|\frac{\frac{(2x)^{n+1}}{3(n+1)+1}}{\frac{(2x)^n}{3n+1}}\right| = \lim_{n\to\infty} \left|\frac{(2x)^{n+1}}{3(n+1)+1} \cdot \frac{3n+1}{(2x)^n}\right| = |2x| \lim_{n\to\infty} \frac{3n+1}{3n+4} = 2|x|$$

Thus, imposing the restriction that the outcome be less than 1, we infer that the x-values for which the series converges satisfy the inequality $2|x| < 1$, or equivalently $|x| < \frac{1}{2}$. So, the interval of convergence is $\left(-\frac{1}{2}, \frac{1}{2}\right)$.

498. This is an incorrect application of the nth term test for divergence. Applying the limit comparison test with divergent harmonic series $\sum\limits_{n=1}^{\infty} \frac{1}{n}$, we conclude that since $\lim\limits_{n\to\infty} \frac{\frac{1}{3n+1}}{\frac{1}{n}} = \lim\limits_{n\to\infty} \frac{n}{3n+1} = \frac{1}{3} > 0$, the series $\sum\limits_{n=0}^{\infty} \frac{1}{3n+1}$ also diverges.

499. The squeeze theorem doesn't apply in this situation since the outer sequences $\left\{-1 - \frac{1}{n}\right\}$ and $\left\{1 + \frac{1}{n}\right\}$ do not converge to the same limit. Indeed, they approach -1 and 1, respectively. Therefore, nothing definitive can be concluded about the behavior of $\{c_n\}$. For instance, it could be the case that $c_n = -1$ for every n, so the sequence $\{c_n\}$ would converge. Or it could be the case that $c_n = (-1)^n$ for every n, so the sequence $\{c_n\}$ would diverge.

500. Continuity does not imply differentiability. Rather, the opposite is true. The function $f(x) = |x - 1|$ is continuous at every real number x, but it is not differentiable at $x = 1$ because its graph has a sharp corner at that point.

501. The fundamental theorem of calculus does not apply to the integral $\int_{-2}^{1} \frac{1}{x^2} dx$, because the integrand is discontinuous at $x = 0$, which lies in the interval of integration. This integral is *improper*, and a limiting scheme must be used to compute it. Given the topic coverage included within this book, it suffices to say that the theory, as presented, does not apply.

Posttest

Now that you have worked through the problems in all of the chapters, it is time to show off your new skills. Take the posttest to see how much your calculus skills have improved. The posttest has 20 multiple-choice, true/false, and computation questions covering all of single-variable calculus. While the format of the posttest is similar to the pretest, the questions are different.

After you complete the posttest, check your answers using the answer key at the end of this section. If you still have weak areas, go back and work through the applicable problems again.

Questions

1. Compute: $\lim\limits_{h \to 0} \frac{3(x+h)^3 - 3x^3}{h}$

2. $\lim\limits_{x \to 2\pi} \frac{\cos x - 1}{\cos x + 1} = \underline{\hspace{2cm}}$
   a. 0
   b. The solution does not exist.
   c. -1
   d. -3

3. Compute the derivative of $f(x) = (3x^7 - 2x^5 - 3x - 2)^{-15}$.

4. On which of the following sets is the graph of $f(x) = \arctan(2x)$ concave down?
   a. $\mathbb{R}$
   b. $\left(-\frac{\pi}{8}, \frac{\pi}{8}\right)$
   c. $(0, \infty)$
   d. $(-\infty, 0)$

5. A painter has enough paint to cover 800 square feet of area. What is the largest square-bottomed box that could be painted (including the top, bottom, and all sides)?

6. Assume that $\int_0^6 f(x)\,dx = 15$, $\int_6^7 f(x)\,dx = -10$, and $\int_7^{11} f(x)\,dx = 4$. Compute $\int_0^7 -2f(x)\,dx$.

7. Compute: $\int \frac{1 + \ln x}{\cos^2(x \ln x)}\,dx$

8. Compute: $\int x \cos(\pi x)\,dx$

9. Compute: $\int \left(-\csc^2 x + 2\tan x \sec x\right)dx$

10. Compute: $\int \frac{2x}{1 + 9x^4}\,dx$

11. Compute: $\int \cos(\pi x) \cdot \ln(\sin(\pi x))\,dx$

12. Compute: $\int \frac{2x - 7}{(x-3)(x+5)}\,dx$

13. The length of the portion of the curve $y = \int_{3}^{x} \sqrt{9t^2 - 1}\, dt$ starting at $x = 2$ and ending at $x = 3$ is equal to which of the following?

    a. $\sqrt{57}$ units

    b. 3 units

    c. 7.5 units

    d. none of the above

14. The general solution of the differential equation $\frac{dy}{dx} = -\frac{x}{x+1}$ is which of the following?

    a. $y(x) = -x + \frac{1}{(x+1)^2} + C$, where C is a real number

    b. $y(x) = x - \frac{1}{2}x^2 + C$, where C is a real number

    c. $y(x) = -\frac{x^2}{x^2 + x} + C$, where C is a real number

    d. $y(x) = -x + \ln|x+1| + C$, where C is a real number

15. Which of the following is an accurate characterization of the series $\sum_{n=1}^{\infty} \left[2 + \left(\frac{1}{5} \right)^{n-1} \right]$?

    a. The series is convergent with sum 2.25.

    b. The series is convergent with sum 3.

    c. The series is convergent with sum 2.

    d. The series is divergent.

16. Which of the following is the interval of convergence for $\sum_{n=0}^{\infty} \frac{(2n+1)^2 (x+1)^{2n}}{3^{2n-1}}$?

    a. $(-10,8)$

    b. $[-10,8]$

    c. $[-4,2]$

    d. $(-4,2)$

17. Compute with the help of a known power series: $\int e^{x^4}\, dx$

18. The graph of the parametrically defined curve $x = -3 + \sin^2 t$, $y = -1 + \cos^2 t$, $0 \le t \le \frac{\pi}{2}$, is a portion of a(n)

    a. circle.

    b. line.

    c. parabola.

    d. ellipse.

19. Determine the slope of the tangent line to the polar curve $r = 3 - 2\cos\theta$, $0 \leq \theta \leq 2\pi$, when $\theta = \frac{9\pi}{4}$.

20. Which of the following integrals represents the volume of a right circular cylinder with base radius R and height $4H$?

   a. $\pi \int_{-2H}^{2H} R^2 \, dx$

   b. $2\pi \int_{-2H}^{2H} Rx \, dx$

   c. $\pi \int_{0}^{R} (4H)^2 \, dx$

   d. $2\pi \int_{0}^{R} 4Hx \, dx$

Answers

1. To compute this limit, first simplify the expression $(x + h)^3$, as follows:
$$(x+h)^3 = (x+h)^2(x+h) = \left(x^2 + 2hx + h^2\right)(x+h)$$
$$= x^3 + 2hx^2 + h^2 x + x^2 h + 2h^2 x + h^3 = x^3 + 3hx^2 + 3h^2 x + h^3$$
Hence, the original problem is equivalent to
$$\lim_{h\to 0} \frac{3\left(x^3 + 3hx^2 + 3h^2 x + h^3\right) - 3x^3}{h}$$
Now, simplify the numerator, cancel factors that are common to both the numerator and the denominator, and then substitute $h = 0$ into the simplified expression, as follows:
$$\lim_{h\to 0} \frac{3(x+h)^3 - 3x^3}{h} = \lim_{h\to 0} \frac{3\left(x^3 + 3hx^2 + 3h^2 x + h^3\right) - 3x^3}{h}$$
$$= \lim_{h\to 0} \frac{3x^3 + 9hx^2 + 9h^2 x + 3h^3 - 3x^3}{h}$$
$$= \lim_{h\to 0} \frac{h\left(9x^2 + 9hx + 3h^2\right)}{h} = \lim_{h\to 0}\left(9x^2 + 9hx + 3h^2\right) = 9x^2$$

2. a. Substituting $x = 2\pi$ directly into the expression $\frac{\cos x - 1}{\cos x + 1}$ yields
$$\frac{\cos 2\pi - 1}{\cos 2\pi + 1} = \frac{1-1}{1+1} = \frac{0}{2} = 0$$
Thus, $\lim_{x\to 2\pi} \frac{\cos x - 1}{\cos x + 1} = 0$.

3. Applying the chain rule yields
$$f'(x) = -15\left(3x^7 - 2x^5 - 3x - 2\right)^{-16} \cdot \left(3x^7 - 2x^5 - 3x - 2\right)'$$
$$= -15\left(3x^7 - 2x^5 - 3x - 2\right)^{-16} \cdot \left(21x^6 - 10x^4 - 3\right)$$
$$= \frac{-15\left(21x^6 - 10x^4 - 3\right)}{\left(3x^7 - 2x^5 - 3x - 2\right)^{16}}$$

4. c. The graph of $f(x) = \arctan(2x)$ is concave down on precisely those intervals where $f''(x) < 0$. We compute the first and second derivatives of f, as follows:
$$f'(x) = \frac{1}{1+(2x)^2}\cdot(2x)' = \frac{2}{1+4x^2} = 2\left(1+4x^2\right)^{-1}$$
$$f''(x) = -2\left(1+4x^2\right)^{-2}\left(1+4x^2\right)' = -2\left(1+4x^2\right)^{-2}(8x) = \frac{-16x}{\left(1+4x^2\right)^2}$$
Since the denominator of $f''(x)$ is always nonnegative, the only x-values for which $f''(x) < 0$ are those for which the numerator is negative. This happens for only those x-values in the interval $(0, \infty)$.

5. Since the box has a square bottom, its length and width can both be x, while its height is y. Thus the volume is given by Volume $= x^2 y$ and the surface area is Area $= x^2 + 4xy + x^2$ (the top, the four sides, and the bottom). Since Area $= 2x^2 + 4xy = 800$, it follows that the height $y = \frac{800 - 2x^2}{4x} = \frac{200}{x} - \frac{x}{2}$. Thus, the volume function is given by

$$\text{Volume}(x) = x^2 y = x^2 \left(\frac{200}{x} - \frac{x}{2} \right) = 200x - \frac{1}{2}x^3$$

Applying the first derivative test requires that we compute the first derivative, as follows:

$$\text{Volume}'(x) = 200 - \frac{3}{2}x^2$$

Observe that Volume$'(x)$ is defined at all nonzero real numbers, and the only x-value that makes it equal zero is when $x^2 = \frac{400}{3}$. Since

negative lengths are impossible, this is zero only when $x = \frac{20}{\sqrt{3}} = \frac{20\sqrt{3}}{3}$.

Since the sign of Volume$'(x)$ changes from $+$ to $-$ at this x-value, we conclude that Volume(x) has a local maximum at $x = \frac{20\sqrt{3}}{3}$ feet. The

corresponding height is $y = \frac{200}{\frac{20\sqrt{3}}{3}} - \frac{\frac{20\sqrt{3}}{3}}{2} = \frac{30\sqrt{3}}{3} - \frac{10\sqrt{3}}{3} = \frac{20\sqrt{3}}{3}$ feet, so

we conclude that the largest box that could be painted is a cube with all sides of length $\frac{20\sqrt{3}}{3}$ feet.

6. Using interval additivity and linearity, we see that

$$\int_0^7 -2f(x)\,dx = -2\int_0^7 f(x)\,dx = -2\left[\int_0^6 f(x)\,dx + \int_6^7 f(x)\,dx \right]$$
$$= -2\left[15 + (-10) \right] = -10$$

7. Make the following substitution:

$u = x\ln x$

$du = \left[x\left(\frac{1}{x}\right) + (\ln x)(1) \right]dx = (1 + \ln x)\,dx$

Applying this substitution and computing the resulting indefinite integral yields

$$\int \frac{1 + \ln x}{\cos^2(x\ln x)}\,dx = \int \frac{1}{\cos^2 u}\,du = \int \sec^2 u\,du = \tan u + C$$

Resubstituting $u = x\ln x$ into this expression yields

$$\int \frac{1 + \ln x}{\cos^2(x\ln x)}\,dx = \tan(x\ln x) + C$$

8. Apply the integration by parts formula $\int u\,dv = uv - \int v\,du$ with the following choices of u and v, along with their differentials:

$u = x \qquad dv = \cos(\pi x)dx$

$du = dx \qquad v = \int \cos(\pi x)dx = \frac{1}{\pi}\sin(\pi x)$

Applying the integration by parts formula yields:

$\int x\cos(\pi x)dx = (x)\left(\frac{1}{\pi}\sin(\pi x)\right) - \frac{1}{\pi}\int \sin(\pi x)dx$

$$= \frac{1}{\pi}x\sin(\pi x) - \frac{1}{\pi}\left(-\frac{1}{\pi}\cos(\pi x)\right) + C$$

$$= \frac{1}{\pi}x\sin(\pi x) + \frac{1}{\pi^2}\cos(\pi x) + C$$

Note: Computing both $\int \cos(\pi x)dx$ and $\int \sin(\pi x)dx$ entails using the substitution $z = \pi x,\ \frac{1}{\pi}dz = dx$.

9. $\int\left(-\csc^2 x + 2\tan x \sec x\right)dx = \cot x + 2\sec x + C$

10. First, rewrite the integral in the following equivalent manner:

$\int \frac{2x}{1+9x^4}dx = 2\int \frac{x}{1+\left(3x^2\right)^2}dx$

Make the following substitution:

$u = 3x^2$

$du = 6x\,dx \implies \frac{1}{6}du = x\,dx$

Applying this substitution in the integrand and computing the resulting indefinite integral yields

$\int \frac{2x}{1+9x^4}dx = 2\int \frac{x}{1+\left(3x^2\right)^2}dx = 2\int \frac{1}{1+u^2}\left(\frac{1}{6}\right)du = \frac{1}{3}\int \frac{1}{1+u^2}du = \frac{1}{3}\arctan u + C$

Finally, rewrite the final expression of the preceding equation in terms of the original variable x by resubstituting $u = 3x^2$ to obtain

$\int \frac{2x}{1+9x^4}dx = \frac{1}{3}\arctan\left(3x^2\right) + C$

11. We first apply the substitution technique to simplify the integral. Precisely, make the following substitution:

$z = \sin(\pi x)$

$dz = \pi \cos(\pi x) dx \implies \frac{1}{\pi} dz = \cos(\pi x) dx$

Applying this substitution yields the following equivalent integral:

$\int \cos(\pi x) \cdot \ln(\sin(\pi x)) dx = \frac{1}{\pi} \int \ln z \, dz$

Now, apply the formula for integration by parts $\int u \, dv = uv - \int v \, du$ with the following choices of u and v, along with their differentials:

$u = \ln z \qquad dv = dz$

$du = \frac{1}{z} dz \qquad v = \int dz = z$

Applying the integration by parts formula yields:

$\int \ln z \, dz = z \ln z - \int z \left(\frac{1}{z}\right) dz = z \ln z - \int dz = z \ln z - z + C = z(\ln z - 1) + C$

Substituting this back into the equality obtained from our initial step of applying the substitution technique, and subsequently resubstituting $z = \sin(\pi x)$, yields

$\int \cos(\pi x) \cdot \ln(\sin(\pi x)) dx = \frac{1}{\pi}(z \ln z - z) + C = \frac{1}{\pi} z(\ln z - 1) + C$

$$= \frac{1}{\pi} \sin(\pi x) \cdot (\ln(\sin(\pi x)) - 1) + C$$

12. First, apply the method of partial fraction decomposition to rewrite the integrand in a more readily integrable form. The partial fraction decomposition has the form:

$$\frac{2x-7}{(x-3)(x+5)} = \frac{A}{x-3} + \frac{B}{x+5}$$

To find the coefficients, multiply both sides of the equality by $(x-3)(x+5)$ and gather like terms to obtain

$2x - 7 = A(x+5) + B(x-3)$

$2x - 7 = (A + B)x + (5A - 3B)$

Now, equate corresponding coefficients in the preceding equality to obtain the following system of equations whose unknowns are the coefficients we seek:

$$\begin{cases} A + B = 2 \\ 5A - 3B = -7 \end{cases}$$

Now, solve this system. Multiply the first equation by -5 to obtain $-5A - 5B = -10$, and add this to the second equation and solve for B to obtain:

$-8B = -17 \implies B = \frac{17}{8}$

Substituting this into the first equation yields $A = 2 - \frac{17}{8} = -\frac{1}{8}$. Hence, the partial fraction decomposition becomes:

$$\frac{2x-7}{(x-3)(x+5)} = \frac{-\frac{1}{8}}{x-3} + \frac{\frac{17}{8}}{x+5}$$

We substitute this expression in for the integrand in the original integral, compute each, and simplify, as follows:

$$\int \frac{2x-7}{(x-3)(x+5)}\,dx = \int \left(\frac{-\frac{1}{8}}{x-3} + \frac{\frac{17}{8}}{x+5} \right) dx = -\frac{1}{8}\int \frac{1}{x-3}\,dx + \frac{17}{8}\int \frac{1}{x+5}\,dx$$

$$= -\frac{1}{8}\ln|x-3| + \frac{17}{8}\ln|x+5| + C$$

13. c. Since the fundamental theorem of calculus implies that $f'(x) = \sqrt{9x^2 - 1}$, applying the length formula $\int_a^b \sqrt{1 + (f'(x))^2}\,dx$ yields the following:

$$\text{Length} = \int_2^3 \sqrt{1 + \left(\sqrt{9x^2 - 1}\right)^2}\,dx = \int_2^3 \sqrt{9x^2}\,dx = \int_2^3 3x\,dx = \frac{3}{2}x^2 \Big|_2^3$$

$$= \frac{3}{2}x^2 \Big|_2^3 = \frac{3}{2}(9-4) = \frac{15}{2} = 7.5 \text{ units}$$

14. d. Separate variables and integrate both sides to obtain

$$dy = -\frac{x}{x+1}dx \implies \int dy = -\int \frac{x}{x+1}dx$$

The integral on the left side is simply y. Compute the integral on the right side as follows:

$$-\int \frac{x}{x+1}dx = -\int \frac{x+1-1}{x+1}dx = -\int \left[\frac{x+1}{x+1} - \frac{1}{x+1}\right]dx = -\int \left[1 - \frac{1}{x+1}\right]dx$$

$$= -x + \ln|x+1| + C$$

Thus, $y(x) = -x + \ln|x+1| + C$.

15. d. Note that $\lim\limits_{n \to \infty}\left[2 + \left(\frac{1}{5}\right)^{n-1}\right] = 2 + 0 = 2$. Since this value is not zero,

we conclude immediately from the nth term test for divergence that the

series $\sum\limits_{n=1}^{\infty}\left[2 + \left(\frac{1}{5}\right)^{n-1}\right]$ diverges.

16. d. The interval of convergence is determined by applying the ratio test, as follows:

$$\lim_{n\to\infty}\left|\frac{\frac{(2(n+1)+1)^2(x+1)^{2(n+1)}}{3^{2(n+1)-1}}}{\frac{(2n+1)^2(x+1)^{2n}}{3^{2n-1}}}\right| = \lim_{n\to\infty}\left|\frac{(2n+3)^2(x+1)^{2(n+1)}}{3^{2(n+1)-1}}\cdot\frac{3^{2n-1}}{(2n+1)^2(x+1)^{2n}}\right|$$

$$=\lim_{n\to\infty}\left|\frac{(2n+3)^2(x+1)^{2n+2}}{(2n+1)^2\,3^{2n+1}}\cdot\frac{3^{2n-1}}{(x+1)^{2n}}\right|=$$

$$=(x+1)^2\lim_{n\to\infty}\left|\frac{(2n+3)^2}{9(2n+1)^2}\right|=(x+1)^2\lim_{n\to\infty}\left|\frac{4n^2+12n+9}{36n^2+36n+9}\right|$$

$$=\tfrac{1}{9}(x+1)^2$$

Now, we determine the values of x that make the result of this test <1. This is given by the inequality $\frac{1}{9}(x+1)^2<1$, which is solved as follows:

$$\tfrac{1}{9}(x+1)^2<1$$

$$(x+1)^2<9$$

$$(x+1)^2-9<0$$

$$(x+1-3)(x+1+3)<0$$

$$(x-2)(x+4)<0$$

$$-4<x<2$$

Hence, the power series converges at least for all x in the interval $(-4,2)$.

The endpoints must be checked separately to determine if they should be included in the interval of convergence of the power series $\sum\limits_{n=0}^{\infty}\frac{(2n+1)^2(x+1)^{2n}}{3^{2n-1}}$. First, substituting $x=-4$ results in the series

$$\sum_{n=0}^{\infty}\frac{(2n+1)^2(-4+1)^{2n}}{3^{2n-1}}=\sum_{n=0}^{\infty}\frac{(2n+1)^2(-3)^{2n}}{3^{2n-1}}=\sum_{n=0}^{\infty}\frac{(2n+1)^2\,3^{2n}}{3^{2n-1}}=\sum_{n=0}^{\infty}3(2n+1)^2$$

This series clearly diverges because the nth term does not converge to zero as n goes to infinity; rather, the terms go to infinity. So, $x=-4$ is not included in the interval of convergence. As for $x=2$, we must determine if the series $\sum\limits_{n=0}^{\infty}\frac{(2n+1)^2(2+1)^{2n}}{3^{2n-1}}=\sum\limits_{n=0}^{\infty}3(2n+1)^2$ converges.

Here again, the nth term does not go to zero as n goes to infinity. So, $x=2$ is also not included in the interval of convergence. Hence, we conclude that the interval of convergence is $(-4,2)$.

17. We express the integrand as a power series that can be integrated term by term by using $u = x^4$ in the known power series $e^u = \sum\limits_{n=0}^{\infty} \frac{u^n}{n!}$,

$-\infty < u < \infty$, to obtain $e^{x^4} = \sum\limits_{n=0}^{\infty} \frac{\left(x^4\right)^n}{n!} = \sum\limits_{n=0}^{\infty} \frac{x^{4n}}{n!}$. This formula holds for any real number x. Now, substitute this formula in for the integrand and compute as follows:

$$\int e^{x^4}\, dx = \int \sum\limits_{n=0}^{\infty} \frac{x^{4n}}{n!}\, dx = \sum\limits_{n=0}^{\infty} \frac{1}{n!} \int x^{4n}\, dx = \sum\limits_{n=0}^{\infty} \frac{x^{4n+1} + C}{(4n+1)n!},$$

where C is the constant of integration.

18. b. The given parametric equations are equivalent to $x + 3 = \sin^2 t$,

$y + 1 = \cos^2 t$, $0 \le t \le \frac{\pi}{2}$. We can eliminate the parameter t by using the trigonometric identity $\sin^2 \theta + \cos^2 \theta = \mathbf{1}$. Doing so yields the equivalent Cartesian equation $(x + 3) + (y + 1) = 1$, or equivalently $y = -x - 3$; this is a line. The given curve is the portion of this line starting at the point $(-3, 0)$ (when $t = 0$) and ending at the point $(-2, -1)$ (when $t = \frac{\pi}{2}$).

19. In order to write the equation of the tangent line, we need the slope and the point on the curve that occur when $\theta = \frac{9\pi}{4}$. The general formula for the slope of the tangent line to the polar curve $r = 3 - 2\cos\theta$, $0 \le \theta \le 2\pi$, is given by

$$\frac{dy}{dx} = \frac{\frac{dr}{d\theta}\sin\theta + r\cos\theta}{\frac{dr}{d\theta}\cos\theta - r\sin\theta} = \frac{(2\sin\theta)\sin\theta + (3 - 2\cos\theta)\cos\theta}{(2\sin\theta)\cos\theta - (3 - 2\cos\theta)\sin\theta}$$

Evaluating this expression when $\theta = \frac{9\pi}{4}$ yields the slope of the specific tangent line of interest, as follows:

$$\frac{dy}{dx} = \frac{2\sin\frac{\pi}{4}\sin\frac{\pi}{4} + \left(3 - 2\cos\frac{\pi}{4}\right)\cos\frac{\pi}{4}}{2\sin\frac{\pi}{4}\cos\frac{\pi}{4} - \left(3 - 2\cos\frac{\pi}{4}\right)\sin\frac{\pi}{4}} = \frac{2\left(\frac{\sqrt{2}}{2}\right)\left(\frac{\sqrt{2}}{2}\right) + \left(3 - 2\cdot\frac{\sqrt{2}}{2}\right)\frac{\sqrt{2}}{2}}{2\left(\frac{\sqrt{2}}{2}\right)\left(\frac{\sqrt{2}}{2}\right) - \left(3 - 2\cdot\frac{\sqrt{2}}{2}\right)\frac{\sqrt{2}}{2}}$$

$$= \frac{1 + (3 - \sqrt{2})\left(\frac{\sqrt{2}}{2}\right)}{1 - (3 - \sqrt{2})\left(\frac{\sqrt{2}}{2}\right)} = \frac{\frac{2 + 3\sqrt{2} - 2}{2}}{\frac{2 - 3\sqrt{2} + 2}{2}} = \frac{3\sqrt{2}}{4 - 3\sqrt{2}}$$

Next, note that the polar coordinates of the point of tangency (i.e., the point on the curve when $\theta = \frac{9\pi}{4}$) are $r = 3 - 2\cos\frac{9\pi}{4} = 3 - 2\left(\frac{\sqrt{2}}{2}\right) = 3 - \sqrt{2}$.

The corresponding Cartesian coordinates are obtained using the transformation equations $x = r\cos\theta$, $y = r\sin\theta$, as follows:

$$x = r\cos\theta = \left(3 - \sqrt{2}\right)\cos\frac{9\pi}{4} = \left(3 - \sqrt{2}\right)\frac{\sqrt{2}}{2} = \frac{3\sqrt{2}}{2} - 1$$

$$y = r\sin\theta = \left(3 - \sqrt{2}\right)\sin\frac{9\pi}{4} = \left(3 - \sqrt{2}\right)\frac{\sqrt{2}}{2} = \frac{3\sqrt{2}}{2} - 1$$

Summarizing, the slope of the tangent line is $m = \frac{3\sqrt{2}}{4 - 3\sqrt{2}}$ and a point through which the line passes is $\left(\frac{3\sqrt{2}}{2} - 1, \frac{3\sqrt{2}}{2} - 1\right)$. Hence, using the point-slope formula for the equation of a line, we conclude that the equation of the desired tangent line is given by

$$y - \left(\frac{3\sqrt{2}}{2} - 1\right) = \frac{3\sqrt{2}}{4 - 3\sqrt{2}}\left(x - \left(\frac{3\sqrt{2}}{2} - 1\right)\right).$$

20. d. Consider the region bounded by $y = 2H$, $y = -2H$, $x = 0$, and $x = R$: If this region is revolved around the y-axis, a right circular cylinder with base radius R and height $4H$ is formed. Using the method of cylindrical shells, the volume of this solid is given by the integral $2\pi\int_0^R 4Hx\,dx$.

Appendix

Key Calculus Formulas and Theorems

Theorem: One-Sided Limits

- If $\lim\limits_{x \to a^-} f(x) \neq \lim\limits_{x \to a^+} f(x)$ (or at least one of them does not exist), then $\lim\limits_{x \to a} f(x)$ does not exist.

- If $\lim\limits_{x \to a^-} f(x) = \lim\limits_{x \to a^+} f(x) = L$, then $\lim\limits_{x \to a} f(x) = L$.

Arithmetic of Limits

Let n and K be real numbers and assume f and g are functions that have a limit at c.

Then, we have:

Rule (Symbolically)	**Rule (in Words)**
1. $\lim_{x \to c} K = K$	1. Limit of a constant is the constant.
2. $\lim_{x \to c} x = c$	2. Limit of "x" as x goes to C in C.
3. $\lim_{x \to c} Kf(x) = K \lim_{x \to c} f(x)$	3. Limit of a constant times a function is the constant times the limit of the function.
4. $\lim_{x \to c}\left[f(x) \pm g(x) \right] = \lim_{x \to c} f(x) \pm \lim_{x \to c} g(x)$	4. Limit of sum (or difference) is the sum (or difference) of the limits.
5. $\lim_{x \to c}\left[f(x)g(x) \right] = \left(\lim_{x \to c} f(x) \right) \cdot \left(\lim_{x \to c} g(x) \right)$	5. Limit of a product is the product of the limits.
6. $\lim_{x \to c}\left[\dfrac{f(x)}{g(x)} \right] = \dfrac{\lim_{x \to c} f(x)}{\lim_{x \to c} g(x)}$, provided $\lim_{x \to c} g(x) \neq 0$	6. Limit of a quotient is the quotient of the limits, provided the bottom doesn't go to zero.
7. $\lim_{x \to c}\left[f(x) \right]^n = \left[\lim_{x \to c} f(x) \right]^n$, provided the latter is defined.	7. Limit of a function to a power is the power of the limit of the function.

Differentiation Rules

1. **Power rule:** $(x^r)' = rx^{r-1}$
2. **Factor out constants:** $[cf(x)]' = cf'(x)$
3. **Sum/difference rule:** $[f(x) \pm g(x)]' = f'(x) \pm g'(x)$

4. **Product rule:** $[f(x)g(x)]' = f'(x)g(x) + f(x)g'(x)$

   (For more factors, this extends to $(fgh)' = f'gh + fg'h + fgh'$, and so on.)

5. **Quotient rule:** $\left[\dfrac{f(x)}{g(x)}\right]' = \dfrac{g(x)f'(x) - f(x)g'(x)}{[g(x)]^2}$

6. **Chain rule:** If $\underbrace{y = y(u)}_{y \text{ is some combination of } u}$ where $\underbrace{u = u(x)}_{u,\text{ in turn, is some combination of } x}$, then

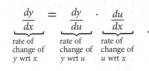

$$\underbrace{\frac{dy}{dx}}_{\substack{\text{rate of} \\ \text{change of} \\ y \text{ wrt } x}} = \underbrace{\frac{dy}{du}}_{\substack{\text{rate of} \\ \text{change of} \\ y \text{ wrt } u}} \cdot \underbrace{\frac{du}{dx}}_{\substack{\text{rate of} \\ \text{change of} \\ u \text{ wrt } x}} .$$

Some Common Derivatives

1. $\dfrac{d}{dx}\sin x = \cos x$

2. $\dfrac{d}{dx}\cos x = -\sin x$

3. $\dfrac{d}{dx}\tan x = \sec^2 x$

4. $\dfrac{d}{dx}\cot x = -\csc^2 x$

5. $\dfrac{d}{dx}\sec x = \sec x \tan x$

6. $\dfrac{d}{dx}\csc x = -\csc x \cot x$

7. $\dfrac{d}{dx}e^x = e^x$

8. $\dfrac{d}{dx}\ln x = \dfrac{1}{x}$

9. $\dfrac{d}{dx}\arcsin x = \dfrac{1}{\sqrt{1-x^2}}$

10. $\dfrac{d}{dx}\arctan x = \dfrac{1}{1 + x^2}$

Properties of the Integral

Let k be a real number and

1. $\displaystyle\int_a^b f(x)\,dx = -\int_b^a f(x)\,dx$

2. **Linearity:** $\displaystyle\int_a^b kf(x)\,dx = k\int_a^b f(x)\,dx,$

   $\displaystyle\int_a^b [f(x) \pm g(x)]\,dx = \int_a^b f(x)\,dx \pm \int_a^b g(x)\,dx$

3. **Interval additivity:** $\int_a^b f(x)\,dx + \int_b^c f(x)\,dx = \int_a^c f(x)\,dx$

4. **Symmetry:**

   i. If f is an even function on $[-a, a]$, then $\int_{-a}^a f(x)\,dx = 2\int_0^a f(x)\,dx$.

   ii. If f is an odd function on $[-a, a]$, then $\int_{-a}^a f(x)\,dx = 0$.

5. **Periodicity:** If f has period p, then $\int_{a+np}^{b+np} f(x)\,dx = \int_a^b f(x)\,dx$, for any integer n.

6. **Antiderivative rule:** $\int_a^b g'(x)\,dx = g(x)\Big|_a^b = g(b) - g(a)$

7. **Fundamental theorem of calculus:** If f is continuous on $[a,b]$, then $\frac{d}{dx}\int_a^x f(t)\,dt = f(x)$. More generally, if u is a differentiable function, then the chain rule yields $\frac{d}{dx}\int_a^{u(x)} f(t)\,dt = f(u(x)) \cdot u'(x)$.

   Second fundamental theorem of calculus: If f is a continuous function and F is an antiderivative of f—that is, $F'(x) = f(x)$—then $\int_a^b f(x)\,dx = F(x)\Big|_a^b = F(b) - F(a)$.

Some Common Antiderivatives

1. $\int \sin x\,dx = -\cos x + C$

2. $\int \cos x\,dx = \sin x + C$

3. $\int \sec^2 x\,dx = \tan x + C$

4. $\int \csc^2 x\,dx = -\cot x + C$

5. $\int \sec x \tan x\,dx = \sec x + C$

6. $\int \csc x \cot x\,dx = -\csc x + C$

7. $\int e^x\,dx = e^x + C$

8. $\int \frac{1}{x}\,dx = \ln|x| + C$

9. $\int x^n\,dx = \frac{x^{n+1}}{n+1} + C,\ n \neq -1$

10. $\int \frac{1}{1+x^2}\,dx = \arctan x + C$

Volumes of Solids of Revolution: Methods of Washers and Cylindrical Shells

Consider a region in the Cartesian of the type displayed in the following figure:

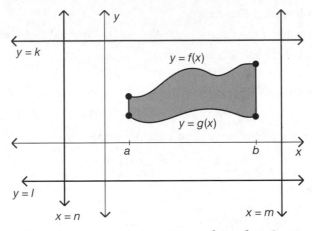

Description of Solid	Integral Used to Compute Volume
1. Revolve region around $y = k$	(Washers)

$$\pi \int_a^b \left[\underbrace{(k - g(x))^2}_{\text{big radius}} - \underbrace{(k - f(x))^2}_{\text{small radius}} \right] dx$$

2. Revolve region around $y = l$ (Washers)

$$\pi \int_a^b \left[\underbrace{(f(x) - l)^2}_{\text{big radius}} - \underbrace{(g(x) - l)^2}_{\text{small radius}} \right] dx$$

3. Revolve region around $x = n$ (Cylindrical shells)

$$2\pi \int_a^b \left[\underbrace{(x - n)}_{\text{radius}} \underbrace{(f(x) - g(x))}_{\text{height}} \right] dx$$

4. Revolve region around $x = m$ (Cylindrical shells)

$$2\pi \int_a^b \left[\underbrace{(m - x)}_{\text{radius}} \underbrace{(f(x) - g(x))}_{\text{height}} \right] dx$$

Length of a Planar Curve

The length of a smooth curve described by a differentiable function $y = f(x)$, $a \le x \le b$, is given by the integral $\int_a^b \sqrt{1 + \left(f'(x)\right)^2}\, dx$.

Average Value of a Function

The average value of the function $y = f(x)$ on the interval $[a,b]$ is given by the integral $\frac{1}{b-a}\int_a^b f(x)\, dx$.

Centers of Mass

The center of mass $\left(\overline{x}, \overline{y}\right)$ of the planar region illustrated in the diagram provided in the earlier brief section on volumes is given by:

$$\overline{x} = \frac{\int_a^b x\, f(x)\, dx}{\text{area of region}}, \quad \overline{y} = \frac{\frac{1}{2}\int_a^b [f(x)]^2\, dx}{\text{area of region}}$$

Convergence of a Sequence

A real-valued sequence $\{x_n\}$ is said to be *convergent* to l if as n gets larger and larger, $|x_n - l|$ gets arbitrarily close to 0. In such case, we write $\lim\limits_{n \to \infty} x_n = l$ (or simply $x_n \to l$). If there is no such target l for which $x_n \to l$, we say that $\{x_n\}$ is *divergent*.

Definition of Infinite Series

- Let $\{a_k\}$ be a sequence of real numbers. Now, define a new sequence of *partial sums* $\{S_n\}$ by $S_n = \sum\limits_{k=1}^{n} a_k$. Then the pair $\left(\{a_n\}, \{S_n\}\right)$ is called an *infinite series*. It is customary to write $\sum\limits_{k=1}^{\infty} a_k$ to represent such an infinite series.

- An infinite series is *convergent* to L $(< \infty)$ if there exists $\lim\limits_{n \to \infty} S_n$. In such case, we write $\sum\limits_{k=1}^{\infty} a_k = L$.

- If either $\lim\limits_{n \to \infty} S_n$ does not exist or $\lim\limits_{n \to \infty} S_n = \pm\infty$, then we say the infinite series is *divergent*.

Geometric Series

A *geometric series* is one of the form $\sum\limits_{k=1}^{\infty} a r^{k-1}$, where $a \neq 0$. Such a series converges (with sum $\frac{a}{1-r}$) if and only if $|r| < 1$.

p-Series

The series $\displaystyle\sum_{n=1}^{\infty} \frac{1}{n^p}$ converges if and only if $p > 1$. (If $p = 1$, the series is called the *harmonic series*.)

Linearity of Convergent Series

Suppose that $\displaystyle\sum_{k=1}^{\infty} a_k$ and $\displaystyle\sum_{k=1}^{\infty} b_k$ *both converge* and let $c \in \mathbb{R}$. Then,

- $\displaystyle\sum_{k=1}^{\infty} c\,a_k$ converges and $\displaystyle = c \sum_{k=1}^{\infty} a_k$.

- $\displaystyle\sum_{k=1}^{\infty} \left(a_k \pm b_k\right)$ converges and $\displaystyle = \sum_{k=1}^{\infty} a_k \pm \sum_{k=1}^{\infty} b_k$.

The *n*th Term Test for Divergence

If $\displaystyle\lim_{k \to \infty} a_k \neq 0$, then $\displaystyle\sum_{k=1}^{\infty} a_k$ diverges.

Ratio Test

Consider the series $\displaystyle\sum_{k=1}^{\infty} a_k$ and let $\displaystyle\lim_{k \to \infty}\left| \frac{a_{k+1}}{a_k} \right| = \rho$.

- If $\rho < 1$, then $\displaystyle\sum_{k=1}^{\infty} a_k$ converges (absolutely).

- If $\rho > 1$, then $\displaystyle\sum_{k=1}^{\infty} a_k$ diverges.

- If $\rho = 1$, then the test is inconclusive.

Limit Comparison Test

Suppose that $\displaystyle\sum_{k=1}^{\infty} a_k$ and $\displaystyle\sum_{k=1}^{\infty} b_k$ are given series. If $\displaystyle\lim_{k \to \infty} \frac{a_k}{b_k} > 0$, then $\displaystyle\sum_{k=1}^{\infty} a_k$ converges (or diverges) if and only if $\displaystyle\sum_{k=1}^{\infty} b_k$ converges (or diverges).

Ordinary Comparison Test

Suppose that $\displaystyle\sum_{k=1}^{\infty} a_k$ and $\displaystyle\sum_{k=1}^{\infty} b_k$ are given series and that $a_k \leq b_k$, for all values of k larger than some integer N. Then,

- If $\displaystyle\sum_{k=1}^{\infty} b_k$ converges, then $\displaystyle\sum_{k=1}^{\infty} a_k$ converges.

- If $\displaystyle\sum_{k=1}^{\infty} a_k$ diverges, then $\displaystyle\sum_{k=1}^{\infty} b_k$ diverges.

Alternating Series Test

Consider the series $\sum_{k=1}^{\infty} (-1)^k a_k$, where $a_k \geq 0$ for all k. If the sequence $\{a_k\}$ decreases to zero, then $\sum_{k=1}^{\infty} (-1)^k a_k$ converges.

Taylor's Theorem

Suppose f is a function possessing derivatives of all orders. Then, $f(x) = \sum_{n=0}^{\infty} \frac{f^{(n)}(c)}{n!} (x-c)^n$ on $(c-R, c+R)$ if and only if

$$\lim_{n \to \infty} \frac{f^{(n+1)}(\theta)}{(n+1)!} (\theta - c)^{n+1} = 0,$$

$$\underbrace{\text{for all } \theta \in (c - R, c + R)}_{\substack{\text{The error in the approximation can be made arbitrarily} \\ \text{small uniformly on the interval of convergence.}}}$$

Furthermore, such a representation is unique.

Common Power Series Representations

The following are the five power series representations for some common elementary functions, together with their intervals of convergence.

1. $e^u = \sum_{n=0}^{\infty} \frac{u^n}{n!}, \ -\infty < u < \infty$

2. $\sin u = \sum_{n=0}^{\infty} \frac{(-1)^n u^{2n+1}}{(2n+1)!}, \ -\infty < u < \infty$

3. $\cos u = \sum_{n=0}^{\infty} \frac{(-1)^n u^{2n}}{(2n)!}, \ -\infty < u < \infty$

4. $\frac{1}{1-u} = \sum_{n=0}^{\infty} u^n, \ -1 < u < 1$

5. $\arctan u = \sum_{n=0}^{\infty} \frac{(-1)^n u^{2n+1}}{2n+1}, \ -1 < u < 1$

Converting between Polar Coordinates and Cartesian Coordinates

Suppose that the Cartesian point (x,y) and the polar point (r, θ) describe the same position in the Cartesian plane. Here, $0 \leq r < \infty$, $0 \leq \theta \leq 2\pi$, and x and y are real numbers. The following equations relate x and y to r and θ:

- For converting from Cartesian to polar, use $x^2 + y^2 = r^2$, $\tan \theta = \frac{y}{x}$.
- For converting from polar to Cartesian, use $x = r\cos\theta$, $y = r\sin\theta$.

Polar Calculus

Suppose that $r = f(\theta)$ is a polar function.

- If f is differentiable, then the slope of the tangent line to the curve $r = f(\theta)$ is given by

$$\frac{dy}{dx} = \frac{\frac{dr}{d\theta}\sin\theta + r\cos\theta}{\frac{dr}{d\theta}\cos\theta - r\sin\theta}$$

- The area enclosed by the graph of $r = f(\theta)$ for $a \leq \theta \leq b$ is given by $\frac{1}{2}\int_a^b \left[f(\theta)\right]^2 d\theta$.

Calculus of Parametrically Defined Functions

Consider a parametrically defined function given by $x = x(t)$, $y = y(t)$, $a \leq t \leq b$.

- If x and y are differentiable, then the slope of the tangent line to the curve is given by

$$\frac{dy}{dx} = \frac{\frac{dy}{dt}}{\frac{dx}{dt}}$$

- If x and y are differentiable, then the length of the curve is given by

$$\int_a^b \sqrt{\left(\frac{dx}{dt}\right)^2 + \left(\frac{dy}{dt}\right)^2}\, dt.$$

- The area enclosed by the curve on the given interval is given by $\int_a^b y(t) \cdot \frac{dx}{dt} dt.$

Additional Online Practice

Whether you need help building basic skills or preparing for an exam, visit the LearningExpress Practice Center! On this site, you can access additional practice materials. Using the code below, you'll be able to log in and access additional online calculus practice. This online practice will also provide you with:

- **Immediate scoring**
- **Detailed answer explanations**
- **Personalized recommendations for further practice and study**

Log in to the LearningExpress Practice Center by using URL: **www.learnatest.com/practice**

This is your Access Code: **7656**

Follow the steps online to redeem your access code. After you've used your access code to register with the site, you will be prompted to create a username and password. For easy reference, record them here:

Username: _____ **Password:** _____

With your username and password, you can log in and access the additional practice material. If you have any questions or problems, please contact LearningExpress customer service at 1-800-295-9556 ext. 2, or e-mail us at **customerservice@learningexpressllc.com**.

Notes

Notes